access to geography

RURAL SETTLEMENT *and* THE URBAN IMPACT *on the* COUNTRYSIDE

Michael Hill

Hodder & Stoughton

A MEMBER OF THE HODDER HEADLINE GROUP

To my three greatest sources of inspiration for this book
a Somerset childhood, studying under BKR and
living in Italy for many years

Acknowledgements

My thanks to: Melvyn and Diana for updating the information on Somerset, Geraldine for making the survey of Cartmel possible, and Martyn for his DEFRA expertise and input.

The publishers would like to thank the following individuals, institutions and companies for permission to reproduce copyright illustrations in this book:

AKG London/British Library: page 130; AKG London/Erich Lessing: page 133; Peter Brookes, *The Times* and Cartoon Study Centre, University of Kent, Canterbury page 137; Ronald Grant Archive: page 135; Hawaii University Press: page 30.

The publishers would also like to thank the following for permission to reproduce material in this book:
Hawaii University Press for an extract from *Chinese Landscapes: The Village as Place* by R Knapp (1992) used on page 29; *The Hindu* (www.thehindu.com) for an extract from their publication dated 29 December 2002, used on page 101; Profile Books for an extract from *Remaking the Landscape* by J Jenkins (2002) used on page 114; Routledge for extracts from *Landscapes of Settlement* by BK Roberts (1996) used on page 23, *Contested Countryside Cultures* by P Cloke and J Little (1997) used on pages 135 and 137.

Every effort has been made to trace and acknowledge ownership of copyright. The publishers will be glad to make suitable arrangements with any copyright holders whom it has not been possible to contact.

Orders: please contact Bookpoint Ltd, 130 Milton Park, Abingdon, Oxon OX14 4SB. Telephone: (44) 01235 827720. Fax: (44) 01235 400454. Lines are open from 9.00–6.00, Monday to Saturday, with a 24 hour message answering service. You can also order through our website www.hodderheadline.co.uk.

British Library Cataloguing in Publication Data
A catalogue record for this title is available from the British Library

ISBN 0 340 80028 3

First Published 2003
Impression number 10 9 8 7 6 5 4 3 2 1
Year 2009 2008 2007 2006 2005 2004 2003

Cover photo: the village of Scanno, Italy (by the author)

Produced by Gray Publishing, Tunbridge Wells, Kent
Printed in Great Britain for Hodder & Stoughton Educational, a division of Hodder Headline, 338 Euston Road, London NW1 3BH by Bath Press Ltd.

Contents

1 Themes, Concepts and Classifications

KEY WORDS

Index of rurality: a method of classifying the degree of 'ruralness' of an area.
Nucleated: (settlement) clustered or concentrated, e.g. a village.
Dispersed: (settlements) small-scale settlements spread out over a wide area, e.g. hamlets and farmsteads.
Index of dispersion: a statistical or method of showing how dispersed an area's settlement pattern is.
Rural–urban continuum: the range and ranking of settlement from farmsteads, hamlets and villages through to towns.
Central Place Theory: the arrangement of settlements of different sizes, according to their populations, status and fields of influence, as proposed by Christaller in 1933.
Beale Codes: an index of rurality used in the USA.

In my beginning is my end. In succession
Houses rise and fall, crumble, are extended,
Are removed, destroyed, restored, or in their place
Is an open field, or a factory, or a by-pass.

T.S. Eliot *East Coker*

In most parts of the world rural settlements have had a very long history and the establishment of an individual village, hamlet or farm often pre-dates any written historical record. This is in direct contrast to most urban settlements, the history of which is generally much better documented in one way or another. Rural settlements are too often perceived of as static entities, but in reality are dynamic and subject to change, whether this be through gradual growth and development or as a result of more cataclysmic external influences.

1 Defining Rural Settlement

At its most basic level the word 'settlement' can be defined as 'the physical expression of the human need for shelter'. Indeed, the most simple forms of rural settlement are little more than basic shelters made of materials such as brushwood, grass or animal hides. Such settlements are particularly associated with harsher physical environments, such as deserts and tundra. At a more sophisticated level a rural settlement cannot be regarded as just a form of shelter, but as a whole complex of buildings, the style and layout of which reflect various aspects of the social organisation and the economic activities of their inhabitants.

The word 'rural' comes from the Latin *rus*, meaning countryside. Thus, rural settlement is that which is located in the countryside, as opposed to a broadly urbanised area. Furthermore, rural settlement is generally, but not exclusively, closely associated with the land and the wide variety of human economies that utilise it. In LEDCs, most rural settlements are still largely engaged in pastoral or arable farming or a hunting-and-gathering economy. In MEDCs, by contrast, rural settlements are becoming divorced from the agricultural land within which they are located. Tourism, retirement, second homes and commuting are increasingly changing the nature of rural settlements in the richer countries of the world.

2 The Rural–Urban Continuum

a) The distinctions between rural and urban settlements

Rural settlements are part of an overall settlement hierarchy that ranges from the small isolated house or farm in the countryside through hamlets, villages and towns up to the vast scale of the metropolis and megalopolis. But when does a rural settlement become an urban one? What is the difference between a large village and a small town? What is the population threshold that separates a rural settlement from an urban one? These frequently asked questions are often quite difficult to address with definitive answers. Even after the various criteria have been examined, it is worth bearing in mind what the *United Nations Demographic Yearbook* says on the matter:

> There is no point in the continuum from large agglomerations to small clusters or scattered dwellings where urbanity disappears and rurality begins; the division between urban and rural populations is necessarily arbitrary.

Robinson (1990) is similarly vague in making the distinction between rural and urban:

> In general terms 'rural' has been regarded as referring to populations in areas of low density and to small settlements, but, beyond this, division between rural and urban is highly problematic.

Several different types of criteria may be used to sort out the problem of defining the differences between rural and urban settlement at the large village–small town level; but none of these is can be definitively applied on the global scale because there are such great variations between settlement types from one country to another. Some of the main criteria that can be used to distinguish rural and urban settlement are:

- the population size
- the administrative status

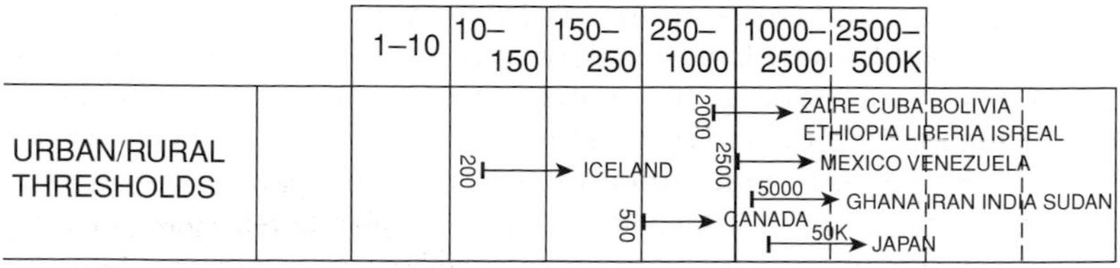

	1–10	10–150	150–250	250–1000	1000–2500	2500–500K
URBAN/RURAL THRESHOLDS		200 → ICELAND	500 → CANADA	2000 → ZAIRE CUBA BOLIVIA ETHIOPIA LIBERIA ISREAL; 2500 → MEXICO VENEZUELA	5000 → GHANA IRAN INDIA SUDAN	50K → JAPAN

Figure 1.1 Rural–urban thresholds (after Roberts, 1996)

- the number and range of services and other functions that the settlement has
- the percentage of the population engaged in agriculture
- the building density
- the building style
- the overall atmosphere or 'feel' of the settlement.

The first of these, population size, illustrates so well how lacking in universal application these criteria are. Figure 1.1 shows the rural–urban population thresholds for selected countries. It is quite clear from this list that countries with low population densities such as Iceland and Canada have much lower thresholds than more densely populated nations such as Japan and India.

Administrative status is equally varied in application. In England and Wales, before the various local government reforms from the 1970s onwards, there were clear distinctions within counties between Municipalities and Urban Districts on the one hand, which had their town councils, and Rural Districts, which had two layers of administration, at the District and Parish levels. This clear distinction has never been the case in some other European countries such as Italy where every council, from that of the Rome or Milan metropolitan area down to villages of less than 100 people, is known as a *commune* and has its own *sindaco* (mayor). In some countries the administrative status is tied to the population threshold mentioned above, so for example in Canada anywhere with a population over 500 will automatically be regarded as a township.

The provision of services and amenities are a relatively accurate guide to whether a settlement has rural or urban status. According to the principles of Christaller's **Central Place Theory**, which will be examined in more detail in Chapter 4, there is a direct link between settlement size and status, and the range of services it provides. In Britain numerous geographers have carried out surveys in different parts of the country to ascertain the relationship between the size of settlements and the number and range of services that they offer. Everson and Fitzgerald did a detailed study of the services provided by both rural and urban settlements in part of East Anglia, using a very wide range of data resources. Carter carried out a detailed analysis of settlements in Pembrokeshire in SW Wales using data from what was then the Dyfed County Council. The findings of such studies as these do, on balance, confirm the principles expected. There are many exceptions to the general rule, however, that make the criterion an imperfect one. Large commuter villages may have a more restricted range of services than purely rural villages of a similar size; the population in a commuter village is able to rely more heavily upon the services provided by the neighbouring city, whereas the population of the purely rural village does not have this advantage and the wider range of services needs to be made available locally. Another exception to the rule is the way in which certain more specialised towns may offer a more restricted range of services than neighbouring rural settlements with smaller populations. The seaside resort with its seasonality and concentration upon catering and leisure services is a good example of this type of variation from the expected pattern. With such difficulties in applying these principles in one country, it can be appreciated that a global application would be almost impossible.

The percentage of people working in agriculture is another criterion that would have a different application in each country. In LEDCs the percentage of people working the land within rural settlements is generally extremely high, whereas in MEDCs it is much lower and on the decline. In Britain and elsewhere in Europe there are now huge variations in the percentage of people working in agriculture between one rural settlement and another. On the one hand, smaller and remoter rural settlements will have a larger portion of their population working on the land, whereas bigger villages close to towns and cities will have much greater concentrations of commuters and retirees.

The building densities, building styles and overall atmosphere or 'feel' of a rural settlement will also vary considerably from one country to another. In northern Europe and North America, there tends to be a greater architectural distinction between rural and urban settlements than in many other parts of the world. The one- or two-storey house, the low density of building and the large amount of green space are typical features of villages or hamlets in Britain, Germany, Scandinavia, New England and the Atlantic Provinces of Canada. By contrast, many rural settlements in Mediterranean Europe, North Africa, the Middle East and the Indian sub-continent have a decidedly 'urban' character, with high building lines and building densities, and very little green space within them. One of the best illustrations of this is the large hilltop village of southern Italy, southern Spain or Greece, which in terms of its character, with three- or four-storey buildings and a central square, is virtually indistinguishable from a small town.

b) Different degrees of rurality

There are considerable variations within the character of rural areas, especially as a result of how close they are located to large urban areas. In 1986 the British geographers Cloke and Edwards put forward the concept of an **index of rurality**, based upon data from the 1981 census. They used 16 different pieces of census data for individual rural areas to draw up their index and included:

- the percentage of the population working in agriculture
- the distance from an urban node with more than 50 000 people
- the percentage of the population involved in commuting
- the percentage of the population over retirement age.

On the basis of this study, Cloke and Edwards identified four different categories of rural and urban environments within Britain.

- **Extremely rural**. This would typically include the remoter rural areas, over 2 hours' drive from a large city, where the farmland would possibly be quite marginal, and a lot of the landscape given over to recreation, as within a National Park. At the same time the area may still be experiencing some rural depopulation from its remoter villages (e.g. the Highlands and Islands of Scotland, and the English Lake District).
- **Intermediate rural**. This zone would be located between 1 and 2 hours' drive from a large city and be comprised mainly of productive farmland, with some more marginal land used for recreation and leisure. The rural settlements are not significantly altered in character by the influence of their closest large city (e.g. much of Herefordshire and Wiltshire).
- **Intermediate non-rural**. This zone lies within a $\frac{1}{2}$–1-hour's drive of a large city and therefore includes the outer part of the main commuter belt. The countryside still is predominantly farmland but there is a significant amount of hobby farming and recreational land use is also very important. The zone has been significantly altered by suburbanisation and many villages have changed in character, and have an old core surrounded by modern housing estates (e.g. north-west Somerset and north-east Cheshire).
- **Extremely non-rural**. This zone extends from the edge of a large city into its immediate surroundings and is generally within a $\frac{1}{2}$–$\frac{3}{4}$-hour's commuting distance. The land use of this **green belt** zone is a mixture of farmland (with a high proportion of hobby farming), recreational land (with a high concentration of golf courses) and protected land such as woods and commons, which are accessible to the general public for recreational use. All towns and villages within this zone have a high percentage of inhabitants who are commuters (e.g. the area outside of Greater London but within the M25, and the zone between Coventry and Birmingham).

This classification can be criticised on a number of counts. It is based upon statistics rather than using more subjective and cultural criteria. Moreover, in the last 20 years since the 1981 census, the impact of the motorways and new technology have tended to reduce actual and psychological distances and to reduce the sense of remoteness of the more peripheral rural areas. All this has contrived to create a greater homogeneity in the character of rural and urban areas alike. These changes within the context of suburbanisation of rural areas will be considered later in Chapter 6.

In Britain today, the Countryside Agency uses a variation on this index of rurality concept in order to differentiate contrasting types of rural areas. The Rural Development Commission Report in 1993, used a slightly different classification based on five categories:

- Metropolitan, which just includes England's seven metropolitan counties
- Urban, which includes all the other major urbanised areas of England
- Coalfield, which includes the semi-urbanised old mining areas such as parts of Derbyshire and County Durham
- Accessible Rural, which includes most shire county areas of central, southern and south-eastern England
- Remote Rural, which includes most of the South-West Peninsula, East Anglia, Lincolnshire, rural NE England and Cumbria.

The inclusion of 'Coalfield' as a separate category is one of the main differences in this classification and it reflects the difficulty of defining mining villages as either urban or rural.

The Council for the Protection of Rural England has mapped what it calls the 'Tranquil Areas of England'. This is in many ways a more useful method of defining rurality as in the twenty-first century it is not just proximity to large cities that cause disturbance to the tranquillity of rural life. The CPRE's definition of a Tranquil Area is based upon numerous pieces of data associated with noise and air pollution and includes the following:

- being over 4 km from a large power station
- being over 3 km from a major city, the most heavily trafficked roads or heavy industries
- being over 1 km away from medium volume roads (i.e. over 10 000 vehicles per day)
- being outside of the flight paths of civil and military airports.

The maps produced by the CPRE show that very great changes have taken place in the last 30 years. In the 1960s 91 880 km^2 of England could be defined as tranquil, but by the 1990s this had declined to 73 012 km^2, a loss of 21% (see Figure 1.2).

In the USA the rural–urban continuum has been codified into ten categories. These enable the analysis of census data, particularly migration trends, and are also important in relation to the funding of social services such as healthcare and education, as well as assessing comparative levels of poverty between town and country. These classifications are known as **Beale Codes**, after Dr Calvin Beale who devised them for the US Department of Agriculture in the 1970s, and

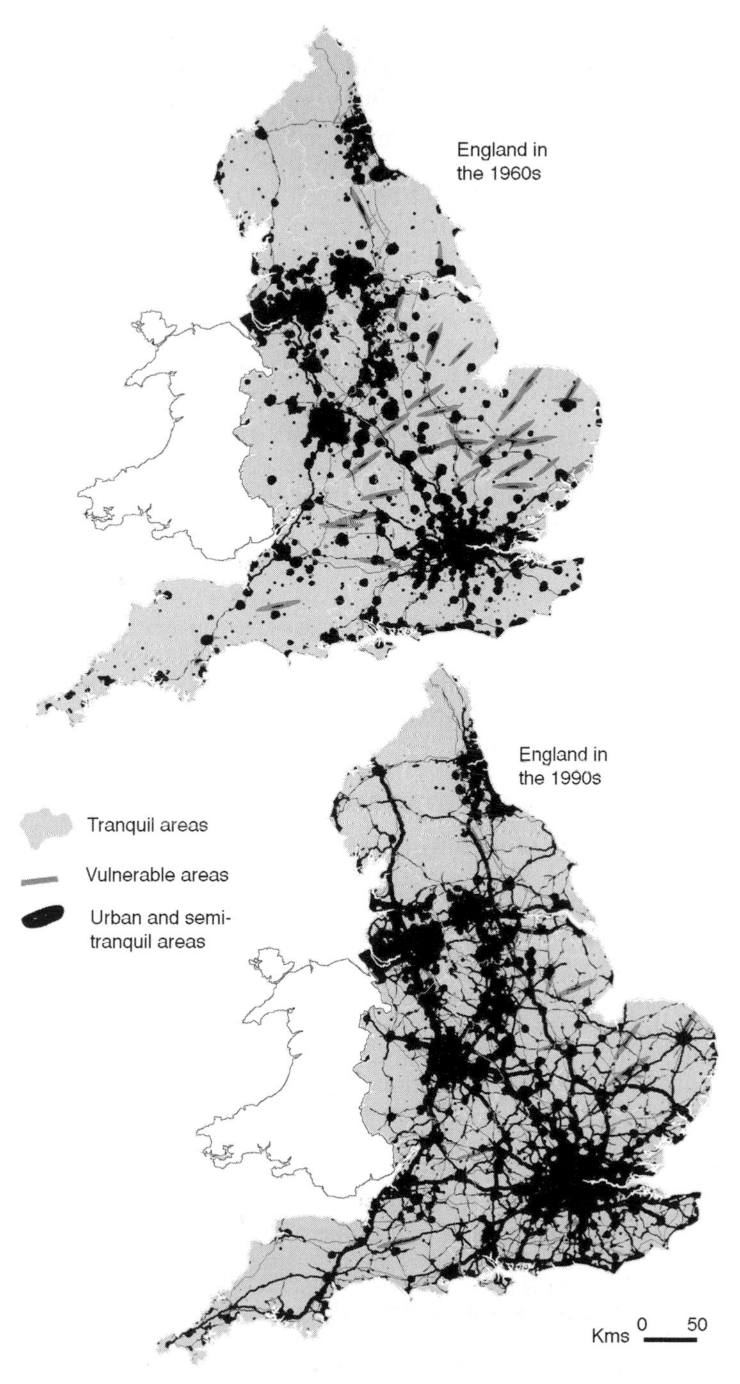

Figure 1.2 Tranquil areas of England

have been revised several times since, the last revision having been made in 1988. The US classification is based upon the county unit and falls into the ten following categories:

- central counties to metropolitan areas with over 1 million people
- fringe counties to metropolitan areas with over 1 million people
- counties with metropolitan areas of 250 000–1 million people
- counties with metropolitan areas with less than 250 000 people
- counties with an urban area with more than 20 000 people adjacent to a metropolis
- counties with an urban area with more than 20 000 people not adjacent to a metropolis
- counties with an urban area of 2500–19 999 people adjacent to a metropolis
- counties with an urban area of 2500–19 999 people not adjacent to a metropolis
- completely rural counties (nowhere with over 2500 people) adjacent to a metropolis
- completely rural counties (nowhere with over 2500 people) not adjacent to a metropolis.

It can be seen quite clearly that this classification can be compared with those used in Britain; the first three categories represent extreme urban, the next three intermediate urban, the next two intermediate rural and the last two categories, extreme rural.

3 Mobility and Permanence

One important way of classifying rural settlement is by its degree of permanence. Categorising in this way shows how both environment and economy influence settlement type. Generally, when a society is highly mobile, this is a response to a severe environment. Deserts and places with severe winters (e.g. the tundra lands and mid-latitude grasslands) have some of the most hostile environments on Earth and their inhabitants have responded accordingly by being very mobile in order to find food. By contrast, areas with an equable climate and rich soils have long supported large, sedentary rural populations. As Roberts (1996) suggests: 'the permanence of a settlement is a function of the power to exploit a restricted environment'.

Schoenauer (1981) places rural settlements and house types into six different categories according to their degree of permanence. He regarded these categories as being like an evolutionary hierarchy of dwelling types of dwellings from the most mobile shelters to the most permanent of buildings.

Schoenauer's classification is as follows:

- **Ephemeral** settlements are those that are only established for a few days. These settlements are associated with highly mobile hunting and gathering societies living in harsh environments, and are particularly found in arid areas. Examples of such settlements are the simple brushwood and grass shelters or *skerms* of the Kung nomads of the Kalahari Desert and the huts of the Arunta Aborigines of Central Australia. These examples have now become almost consigned to history, as both cultural contact with Europeans and deliberate government policies have led to the displacement and re-settlement of both the Kung and the Arunta.

- **Episodic** or **irregular temporary** settlements are those established for a number of weeks at a time. Like ephemeral rural settlements, they are associated with hunting and gathering communities living within difficult or exacting environments, where there is a seasonality to food supply, but it is not always to be found in the same location. Episodic settlements are found in the Arctic, sub-Arctic and harsher continental interiors of the Northern Hemisphere, as well as in the tropical forest environments of low latitudes. The ice-block built winter *igloo* and the summer sealskin tent or *tupic* were, until the mid-twentieth century, the common forms of episodic habitation of the Inuit peoples of the North-Western Territories of Canada; these people were hunters who migrated north and south following the animals they stalked. The large communal homes of various South American rainforest Indians such as the *maloca* of the Erigbaagsta Indians in Brazil and the bell-shaped dwellings of the Makiritare Indians of Venezuela are examples of traditional forms of episodic houses still built today.
- **Periodic** or **regular temporary** settlements are those established for a few weeks at a time and are generally set up in the same place each year. Most of the settlements in this category are associated with nomadic herders in arid and semi-arid parts of the world. These temporary homes are typically tents of one form or another, which can be readily packed up and transported to the next suitable area of grazing land. The *yurts* of the Central Asian nomadic peoples, such as the Mongolians and the Kirgiz, are made of thick pieces of felt placed over a wooden framework and are well adapted to the severe climatic conditions of the semi-arid steppes of that region. The woven goats' hair tents of the Bedouin camel herders of North Africa and the Middle East are also an example of periodic rural settlement. The numbers of people still living in such settlements are in sharp decline for two main reasons. Not only are the younger generations abandoning the traditional way of life and migrating to villages and towns in search of employment, but also various governments, in the name of modernisation, have had deliberate policies of sedentarisation and have forced their nomads to settle down in permanent settlements.
- **Seasonal** settlements are those that are used for several months at a time. They are mostly associated with semi-nomadic peoples who

have mixed agricultural economies involving both livestock rearing and crop cultivation, but who operate in climatic zones that are highly seasonal. Various peoples of the dry savanna lands of Africa, which experience a prolonged dry season and a shorter rainy season, live in these types of settlements. The Nuer tribes people of the Sudan are semi-nomadic cattle herders who supplement their diet of milk, blood and meat with cereals and vegetables grown in the rainy season. The Nuer compounds are *kraals* made of mud built huts surrounding an internal cattle enclosure. Other East African groups that operate similar seasonal economies and live in *kraal* type compounds include the Barabaig of Tanzania and the Masai of Kenya. In recent historic times the Navajo Native American Indians also lived in seasonal settlements in the semi-arid lands of the US South-West. Their seasonal movements were into the mountains in the summer months where they cultivated maize and other crops down to the plains in autumn, winter and spring, where they tended their livestock.

- **Semi-permanent** settlements are those that may be established for many years or even decades. The peoples living in these semi-permanent houses are sedentary farmers who are generally involved in both crop cultivation and animal husbandry. The reasons for these dwellings not being totally permanent vary from place to place, but two of the most common are wear and tear on the buildings themselves and the need to move to another site because of soil exhaustion and land degradation. An economy based upon semi-permanence is hardly practicable in densely populated areas of the world, and is therefore most common in dry savanna lands and semi-arid areas that support sparse populations. Compounds, cluster dwellings and villages throughout the Sahel – the Sahara Desert's southern fringe – fit into this category. The complex cluster dwellings of the Dogon of the Bandiagara Escarpment in central Mali and of the Awuna of northern Ghana, and the compounds of the Guruni of Burkina Faso and Mesakin Quisar of the Sudan are all made of mud with thatched roofs: materials that are easily weathered and frequently need patching up. As these societies do not use sophisticated forms of irrigation to sustain high crop yields, it eventually becomes necessary for the people to move their settlements to a new location. The *pueblo* dwelling Native American Indians of South-West USA are another example of semi-permanent cultivators. In parts of semi-desert plateaux of Arizona and New Mexico the traditional multi-storeyed adobe villages with ladders leading to rooftops are still inhabited today. Before the arrival of Europeans in this region there were thousands of these *pueblos* throughout South-West USA.
- **Permanent** rural settlements are those associated with the most advanced sedentary agricultural societies. These settlements, whether isolated farmsteads or nucleated villages, tend to be

very solidly built and effectively used for several generations. The permanence of such settlements is supported by advanced, stabilised agricultural production that creates a food surplus and can support a growing population. Not only are the family units within these settlements varied in size and type, but they also tend to have well-developed social hierarchies. The huge range of farm, hamlet and village types within Britain and elsewhere in Europe fall into the permanent category, as do most agrarian settlements in the more densely populated parts of the rest of the world.

4 Nucleation and Dispersion

One of the biggest distinctions within the realm of rural settlement is that between **nucleated** and **dispersed** settlements. On the one hand, nucleated rural settlement is clustered and generally concentrated upon one centre, on the other dispersed rural settlement is scattered and may be based upon one or more individual locations. Typically, nucleated rural settlement is that of villages, whereas dispersed is associated with farmsteads and hamlets. Nucleated settlement forms are closely associated with densely populated areas that have highly productive agriculture, whereas dispersed forms are a reflection of lower densities of population and less intensive farming systems. In general terms within Europe, dispersion is more commonly found in the higher and more marginal lands and nucleation on the lowlands, but there are many exceptions to this rule.

In the late nineteenth century the German geographer, Meitzen identified a cultural relationship with these two basic types of settlement. He suggested that the nucleated village (*dorf*) was essentially associated with the Germanic colonisation of central Europe, whereas the isolated farmstead (*einzelhof*), which was its pre-cursor, resulted from the earlier Celtic period of colonisation. Meitzen's findings are useful as a partial explanation of the present settlement patterns in Great Britain. Figure 1.3 shows the pattern of nucleation and dispersion in Britain and it is very clear that many parts of southern and eastern England – the core areas of the Anglo-Saxon settlement – are dominated by large nucleated villages, whereas dispersed farms and hamlets dominate much of Scotland, north-west England, Wales, Devon and Cornwall – the areas that remained home to the Celtic peoples following the Saxon conquest of England. Until the Acts of Enclosure in the eighteenth and nineteenth centuries the patterns would have been even more distinctive. A large proportion of the farmsteads that are associated with the villages of southern and eastern England were established at this time when the huge tracts of common land were divided up between several private landowners.

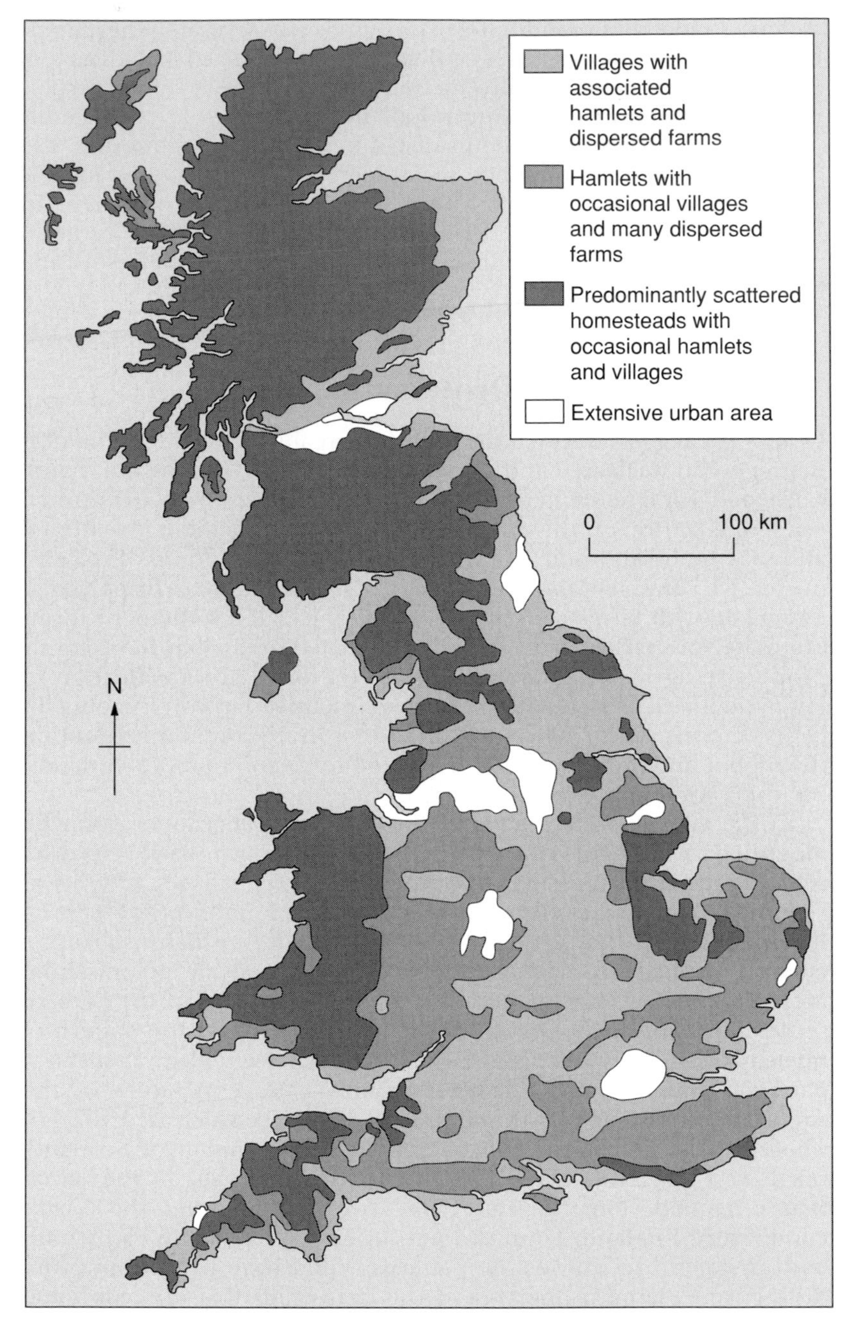

Figure 1.3 Nucleation and dispersion in Britain

Various geographers have attempted to quantify nucleation and dispersion. In the 1930s the French geographer, Demangeon put forward the following formula:

$$K = \frac{E \times N}{T}$$

in this K is the index of dispersion, E is the population of the commune minus that of its main centre, N is the number of settlements and T is the total population of the commune. Using this system, if the K value is between 10 and 50 it would represent a high degree of dispersion, if the value is under $1/10$ then the settlement is highly agglomerated. When Demangeon applied this formula to the Somme Department in northern France, not surprisingly, communes on rich alluvial lowland soils proved to have low K values and high degrees of nucleation, whereas those on higher ground and poorer soils had high K values and were therefore highly dispersed.

During the quantitative revolution in geography in the 1960s, a nearest neighbour statistical formula was applied by the Welsh geographer, Lewis working in Mid Wales. The formula he used was as follows:

$$Rn = 2D\sqrt{N/A}$$

In this case Rn is the pattern of settlement measured, D is the distance to the nearest neighbour, A is the size of the area studied. The Rn values vary from 0 for complete clustering, through 1 for random scattering, to 2.15 for a regularly distributed dispersion.

Roberts (1987) suggests a much simpler formula for distinguishing dispersion from nucleation: the 'hailing distance', or the distance within which farmers could greet each other vocally. He puts this distance at around 150 metres. Following this rule, two farmsteads at a distance of around 200 metres from each other would be a part of a dispersed settlement pattern because the farmers would be beyond the distance they could communicate vocally. By contrast, if the farms were within 100 metres of each other the farmers would be able to 'hail' each other and therefore be regarded as belonging to the same nucleated settlement.

Patterns of nucleation and dispersion are dealt with in more detail in Chapter 4.

5 Continuity and Change

Rural settlements are dynamic entities that are constantly changing and evolving. Population growth at the village level, or even within an individual family leads to the expansion of the very fabric of a farmstead, hamlet or village. Population contraction through disease, famine or war may cause a settlement to shrink in size or even disappear altogether. More cataclysmic events, whether they be natural (e.g.

earthquakes, landslides, epidemics) or human (e.g. invasions, pillaging, expulsions) may lead to the deliberate relocation of a settlement to a less vulnerable place.

Whilst changes may occur in the physical size of a rural settlement through time, it may continue to occupy the same site for centuries or even millennia. This form of continuity reflects the fact that the best village site for one people remains the best for the next. Evidence of the sequent occupancy of one site for a long period of time is often sparse, particularly for places and eras that have no written records; historical geographers and others who have to interpret what changes have occurred have to rely upon archaeological evidence, the interpretation of air photographs and the evidence of place names. Historical geographers have often compared the human landscape of countries that have a long history of settlement as being like a **palimpsest**. Palimpsests were parchment documents dating from Mediaeval times that were reused, after eradicating the underlying earlier script. The landscape in much the same way is the product of a sequence of human modifications superimposed upon it.

In Britain place name evidence is frequently used as a method of distinguishing settlements founded during different historical periods. Each successive wave of colonisers brought with them a different language and culture that has continued to survive in the form of place names. These names often also reveal the importance of certain physical geographical features within the landscape, such as hills, valleys, streams, woods and clearings. Figure 1.4 is a chart showing a selection of place names from the various cultural phases from the Celtic Iron Age period through to the Norman Conquest. Celtic place names still dominate most of Wales and parts of Scotland and Cornwall. Anglo-Saxon names are predominant throughout the southern and western parts of England, and the Midlands, whereas in the east and north of England there is considerable interspersing of Anglo-Saxon and Viking place names. Roman place names are mainly restricted to urban settlements, and Norman names are only found in a very small number of settlements. Although this evidence can be used to distinguish where the different cultures were actively settled, there is nothing to suggest that renaming did not take place. Many of the villages with Saxon names must have been built upon or replaced existing Romano-British (Celtic) settlements, as lowland England was already fairly well populated in Roman times. There are accordingly countless examples of what was most likely to have been the 'Saxonisation' of Celtic place name elements within the names of present-day English villages and hamlets.

Another way in which place name evidence is useful in Britain is in illustrating the continuity of a village name following the subdivision of a parish. In Saxon, Norman and early Mediaeval times, as the population grew and more land was colonised, parishes often became divided into two or more units. Contrasting pairs of place name elements are therefore commonly found throughout Britain, especially in the areas

CELTIC, BRITISH

Element	*Derivation*	*Meaning*
Axe, ESK, USK	isca	water
Quse	us	water
Avon	afon	stream or river
Dee	deva	holy one
Dove	dubo	black
Taw	taw	silent one
Tre-, Trev-,		homestead, village, town
-cet, coed		wood
rhos		moor
barro, bre, brig		various forms or
drum, pen		hill features

ROMAN

Element	*Derivation*	*Meaning*
-caster, chester,		
-cester, etc.	castra	fort or castle
port	portus	harbour
port	porta	gate

ANGLO-SAXON

Element	*Derivation*	*Meaning*
a. Primary settlement (entry phase)		
-ing	ingas	territory of the people of…
-ham	ham	homestead
(-ing is the earlier of these two)		
b. Later primary settlement		
-tn	tun	enclosure
borough, bury,		
burh, byrig		fortified place
bridge		bridge
ford		ford

ANGLO-SAXON (cont'd)

Element	*Derivation*	*Meaning*
c. Early dispersal of daughter settlements		
cot, cote	cote	outlying hut
-field		clearing in wood
-ley	leah	clearing
-stead	stede	place
-stoc, stoke	stoc	daughter settlement
stow		(holy) place
-wick, -wich		outlying hut or shelter, dairy farm, salt pan
d. Later clearing of woodland		
-den		pasture for swine in wood
-jurst, hirst		copse or wooded height
-holt		wood
-weald, wold	wald	wood
-riding, -rod		cleared land
(also hosts of tree names in compound with other elements)		
e. Fen names		
		normally applied to localities;
fen	wet place	if applied to settlements these are generally of
mere, -lake	lake	seventeenth- and eighteenth-century origin
		originally applied to the feature,
-eg, -ey, -ea,	island	but name often transferred to
eig, eu, ea	amidst	a later settlement built in the
	marsh	area

SCANDINAVIAN (DANISH)

Element	*Derivation*	*Meaning*
a. Elements usually applied to settlements		
-toft	topt	homestead or clearing
-by	byr	homestead
-garth	garor	enclosure
-booth	buth	centre for summer pasture
-thorpe	thorp	daughter settlement
thwaite	thveit	clearing
-ergh		outlying hut
b. Elements usually applied to physical features		
-berg, -brig		hill
-ey	ey	island
-force, foss	fors	waterfall
-dale		valley
-gill	geil	ravine
ings	eng	marsh, meadow
beck, slack		stream
tarn		take

Figure 1.4 Place name elements in Britain

of lowland England that had become 'land hungry' in the Middle Ages. These pairs of place name endings may be simply 'North' and 'South', 'East' and 'West', 'Upper' and 'Lower' or 'Great' and 'Little', but there are many more that are much less prosaic. 'Over' and 'Nether' are more traditional Saxon equivalents of 'upper' and 'lower'; 'Magna' and 'Parva' are Latin adjectives for 'great' and 'little'. In many cases the new settlements would take as their suffix their church dedication in order to distinguish them from one another (e.g. St Mary's, St Michael's, St John's); in other cases the family name of the landlord became the place name suffix, and this was particularly common when the Normans arrived thereby introducing names of French noble origin such as Fitzpaine, Neville, Marshall and Beauchamp onto the maps of England. Where the division of a parish led to the establishment of a new settlement upon what was originally uncultivated land, this is often evident through the use of suffixes such as 'Green', 'Common' and 'Heath'.

6 Rural House Types

A great deal of attention has been paid by historical, social and cultural geographers, as well as anthropologists, architects and other specialists, to the different types of buildings to be found in rural areas. Rural houses, farmsteads and farm compounds vary considerably from region to region and from country to country as a result of many different factors that influence their style and layout. These factors include:

- the local climate, especially precipitation and temperature extremes
- local topography, such as steepness of slopes
- building materials
- aspect
- the type of economy practised in the settlement
- political and social structures
- land tenure.

In Britain building material is one of the most important factors influencing rural house type. The country has an extremely complex pattern of geology for its size and this has a profound effect upon vernacular architecture. Certain geological belts provide excellent building stones, whereas others may be limited to brick clays. Although wood and wattle and daub, or cob, were used widely throughout the country right through until the sixteenth century, the seventeenth and eighteenth centuries saw a great deal of rebuilding in villages, particularly in parts of the country that were prosperous through the wool trade.

Now half-timbered rural buildings are the exception rather than the rule. In parts of Essex, Kent, Sussex and the Welsh marches, where good building stones are absent locally, traditional half-timbered cottages and farmsteads are still the dominant house type within individual villages. In Essex and other parts of SE England, weather-boarding is another way in which wood is used in vernacular buildings. In some parts of Britain, the local stone is used as both a walling and roofing material, making houses particularly harmonious with the landscape. In Cornwall, North Wales and the Lake District slate is often used for both walls and roofs, as is Jurassic Limestone in the Cotswolds.

Somerset provides an interesting microcosm for England as it has such a wide variety of building materials within its relatively small area. Villages are given their character by their building materials and in Somerset there are notable differences between the rural settlements made of Carboniferous Limestone and Dolomitic Conglomerate in the Mendip area, Old Red Sandstone on the Quantocks, the Brendons and Exmoor, Blue Lias on and around the Polden Hills, Jurassic Limestone in the east and south of the county and brick down on the Somerset Levels where no local building stone is available.

CASE STUDY: RURAL HOUSE TYPES IN AFRICA

Denyer (1978) in her work *Traditional African Architecture* suggested a taxonomy of African rural house types for which she recognised 32 distinctive house styles. The classification is based upon a number of different characteristics, including:

- overall shape (round, oval, square, rectangular)
- whether or not the structure is free-standing
- what the building material of the walls is (daub, mud-brick, stone, sticks and grass)
- the shape of the roof (hipped, circular, barrel vaulted, domed, flat)
- the roofing material (earth, mud, brick, thatch, fabric)
- whether caves or excavated pits are used as part of the dwelling
- how the individual units are arranged into a village or compound
- the degree of permanence or mobility required of the house or hut type.

Figure 1.5 shows some of the different rural house types to be found in sub-Saharan Africa.

Group A are simple round free-standing huts with mud walls and thatched roofs. These are common throughout the savanna lands of West Africa as well as in the Central African Republic and parts of the Sudan.

Group B are oval or hemispherical, with roofs and walls made of the same materials (mud, skins, brushwood, thatch). These are associated with a certain degree of mobility as found in the cattle herding societies of the dry savanna in countries such as Tanzania, Chad, the Sudan and South Africa.

Group C are known as 'beehive' huts and are generally made entirely of thatched or woven materials such as grasses, reeds and banana leaves. They are found widely through some of the wetter savanna areas of Africa, including Rwanda, Uganda and parts of Nigeria.

Group D are rectangular wattle or daub houses with thatched roofs. Sometimes they are on stilts. Such structures are associated with coastal areas in both western and eastern Africa, as well as villages located on the shores of the great lakes of the Rift Valleys.

Group E are complex mud-brick tower houses two or three storeys high. These are typical of the cattle rearing and millet growing economies of central parts of Benin and Togo as well as much of Burkina Faso.

Group F are well-developed two-storey mud-brick free-standing houses with thatched roofs found in the Tigre region of Ethiopia.

Group G are some of the most sophisticated types of traditional domestic architecture found in Africa. They are houses of one or two storeys made of mud and with flat roofs that are arranged around a central courtyard. These houses sometimes have elaborate decorations on their walls and are typical of the savanna belt of West Africa, in places such as Mali and N Nigeria where they are found in both villages and towns, as in Group H.

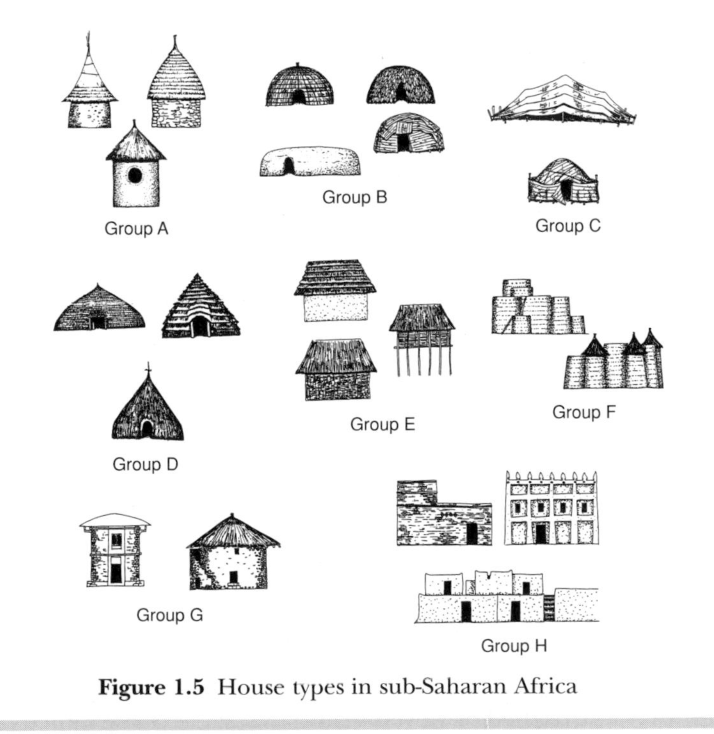

Figure 1.5 House types in sub-Saharan Africa

Questions

1. Examine the validity of the various criteria that may be used to distinguish rural settlements from urban ones.
2. Why is it often difficult to differentiate between a large village and a small town?
3. Explain the relevance of 'dispersion' and 'nucleation' and 'continuity' and 'change' to the study of rural settlements.
4. What is meant by 'different degrees of rurality'? Explain how these may be defined.
5. Outline the different ways in which rural settlements can be classified.

Summary Diagram

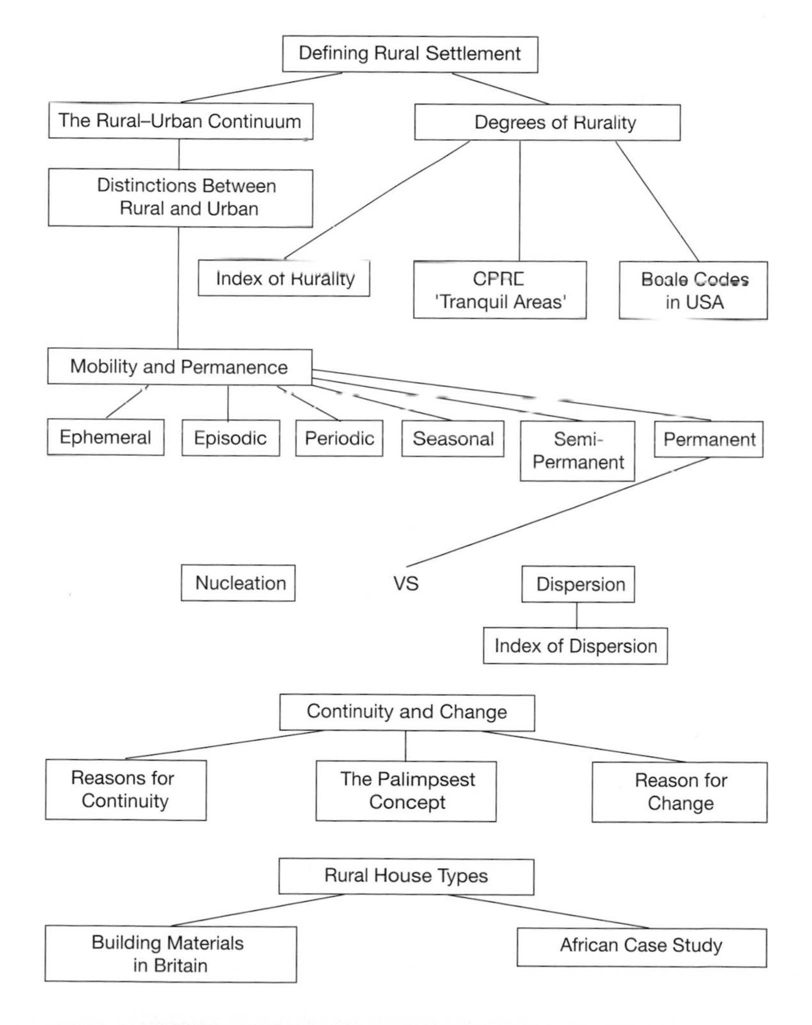

2 Rural Settlement Location

KEY WORDS

Site: the actual area of land upon which a settlement is built.
Position: where a place is located in relation to its surrounding area.
Intrinsic site qualities: natural advantages associated with site.
Extrinsic site qualities: natural advantages associated with position.
Communality: the people's attitude towards the village and their degree of communal organisation.
Geomancy: ancient practices that enabled people to 'read' the landscape and its suitability for settlement.

The wise take pleasure in rivers and lakes,
the virtuous in mountains.

Confucius

1 Introduction

Although each rural settlement location is unique, there are certain underlying principles that influence how a farm or village is positioned according to both what the local environment offers and what the original colonising inhabitants required. How accommodating the physical site is will vary greatly from one climatic or topographical zone of the world to another; there are very different locational determinants in, for example, desert areas, tropical forest zones, the tundra areas and mountainous regions. Even within relatively small areas such as Britain, there are sufficiently great variations in physical geography to make the locations of rural settlements different from one region to another. The locational needs of different human groupings also vary considerably from one part of the world to another because of the wide range of different cultural, social and religious values, and the range of economic activities to be found within human societies.

2 Site and Position

The location of a rural settlement is made up from two distinctive elements, the **site** and the **position** or **situation**. The site is the actual area of land upon which a farm or village is built. Thus, the important considerations of site would include:

- the suitability of land for building upon (whether, for example, it is a rocky, marshy, sandy or clayey site)

- the availability of water (whether or not it is on a surface or underground water supply)
- the actual topography of the site (flat, elevated or, if on a slope, how steep is it?)
- the elevation of the site (how high above sea level?)
- the dryness of the site (is it liable to flood?)
- the aspect of the site (if it is on a slope which direction is it facing?)
- the degree of shelter the relief provides from storms or cold winds.

By contrast to its site, the position or situation of a rural settlement is where it is located in the context of the surrounding area. Factors included in the position of a farm or village would include:

- where it is located in relation to the broader topographical features in the landscape such as mountains, hills, valleys and plains
- how it is connected to natural route ways and lines of communications that have developed along them
- where it is in relation to seas, lakes and other bodies of water
- how it fits into the local settlement hierarchy and how well connected it is to other settlements
- how close it is to various natural resources, such as different soil types, minerals, fuel and building materials.

3 The Main Factors Influencing Rural Settlement Location

Many of the main factors influencing the siting and positioning of rural settlements are listed in the section above, but to understand their significance involves looking at individual places and the minds of those making the decisions of where to locate a settlement. It must be borne in mind that the vast majority of rural settlements in most parts of the world have a long history and therefore their origins are likely to be atavistic. The conditions and needs of the past have to be appreciated in order to understand why farms and villages were established in their particular locations. In an age of piped water supplies and tarmac roads throughout Europe it is easy to regard settlement location as being much more footloose than it really was in the past.

It is also easy in a secular age to diminish the role of the spiritual. In the past, much of the decision making for settlement location was carried out according to arcane practices such as divination, religious rituals or magic rites. The ancient Romans used the doctrine of the *haruspices* when founding new settlements, i.e. the priest examined the liver of a sacrificial animal in order to divine whether a specific spot was suitable for habitation and would give rise to rich harvests. In China there has been a long tradition of the use of *feng shui* principles when locating new villages, and this will be discussed later on in this chapter.

Much of what has been written about rural settlement location has focused down on the northern European cultural realm, where there has been a great tradition of academic research. In the earlier part of the twentieth century, Demangeon working in France and Meitzen in Germany gave significant input into the study of the relationships between settlement and its agricultural context. More recently, British geographers such as Michael Chisholm and Brian Roberts have produced important theories and ideas about rural settlement location.

In 1962 Chisholm published *Rural Settlement and Land Use*, which develops upon the theories of the nineteenth-century German agrarian economist, Johann Heinrich Von Thünen, about the arrangement of land use surrounding agricultural settlements. Von Thünen argued that different activities take place at varying distances from a village or market town, because intensity of agriculture diminishes with distance from market. Chisholm inverted the analysis and came up with a simple model to explain the ideal location of a village in relation to five selected natural resources, shown in Figure 2.1. The setting for the model is the Anglo-Saxon period in England, when a

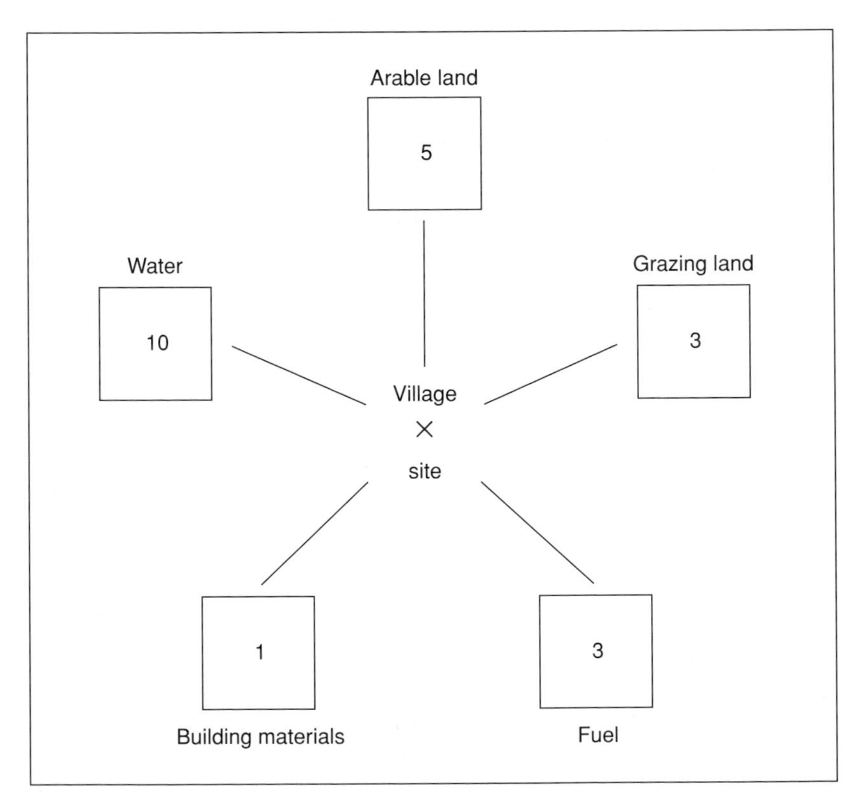

Figure 2.1 Chisholm's model of settlement location

large number of the present-day villages were founded. Chisholm's premise is based on how frequently each of the natural resources is used and how labour intensive the exploitation of each resource is. The numbers represent the mathematical weighting that each of the resources is given according to its relative importance; thus, water supply is awarded twice the importance of arable land and ten times that of building materials. The logic behind this is that water is used daily in large quantities and involves considerable human effort in fetching it. Arable land is visited frequently, especially during certain seasons such as at harvest time, by a large labour force. By contrast, building materials, although bulky, are used very infrequently and the processes involved in working them and using them are not particularly very labour intensive.

Chisholm's model has been used and adapted by numerous geographers as a way of explaining rural settlement location. By providing worked examples based upon theoretical maps with several potential village sites, they enable the calculation of the most ideal location by measuring the distances between each site and resource, and then multiplying the figures by the appropriate weightings; the site with the lowest overall total will represent the most efficient and therefore ideal location.

Roberts has made some of the most detailed contributions to the study of rural settlement in Britain over the last few decades, especially in two works, *The Making of the English Village* (1987) and *Landscapes of Settlement* (1996). In dealing with the location of settlements he makes a clear distinction between **intrinsic** and **extrinsic** qualities of settlement sites (see Figure 2.2). The intrinsic qualities are those more closely associated with a village's site rather than its position. These include water supply, aspect, shelter, flat land, free drainage, local accessibility (i.e. within the village) and the perception of hazards. This last factor was not only connected with problems such as flooding and strong winds, but would also have included supernatural hazards such as demons and ghosts, and the realms within the parish that they inhabited such as dark forests and exposed hillsides. These are all largely physical factors that the village founding ancestors would have had to take carefully into account; in addition to these, Roberts adds two other intrinsic factors that are more social than physical: the need for defence and what he calls 'culturally perceived qualities'. Defensive factors were more important for some settlements than others, depending on whether or not they were located in turbulent parts of the country; these factors were also more important in some stages of history than in others, as in the past there were sequences of invasions from outside interspersed with periods of peace. The cultural qualities of the site are much more difficult to make generalisations about, as they include the role of religion and other cult practices; it is often impossible to know exactly what the cultural values of past societies were. As Roberts (1996) puts it:

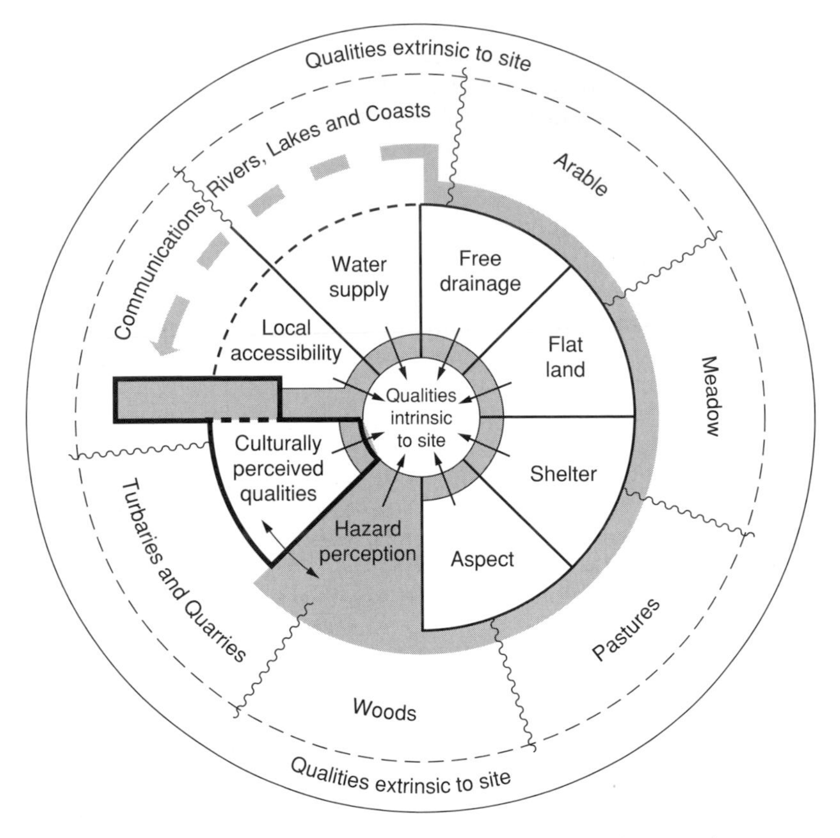

Figure 2.2 Intrinsic and extrinsic qualities in settlement sites (after Roberts, 1987)

> The idea of culturally perceived qualities opens a Pandora's box of questions, for we all see the world and life through our own lifestyles, experiences and beliefs,

Each society that occupied Britain had its holy places. Within a settlement the most important site to the villagers would be that of the place of worship, both in the pre-Christian and the Christian eras. It is not surprising that many Saxon churches were built on the sites of Roman temples; the sites that were sacred to one society would frequently be equally auspicious to the next.

Two important facts must be stressed about the intrinsic settlement sites:

- They were, and are, very rarely just one landscape unit of uniform soil and drainage quality, with a uniform angle of slope. They, in fact, generally have a whole series of slope units of varying land quality providing a variety useful micro-environments.

- They are no longer purely natural environments. Over the millennia, slopes, soil quality and drainage have all been significantly altered by careful tending of the village community in its efforts to make the site more productive and more comfortable.

The extrinsic site qualities are more closely associated with situation than site and involve what the territories surrounding rural settlement have to offer the villagers in terms of resources that can be exploited. The main qualities are therefore arable land meadow and pasture for the main farming activities; woodlands for timber, fuel and foraging; stone for quarrying; turf for fuel or roofing material; valleys and other lines of communication for access to the outside world. In addition to these fairly ubiquitous qualities are others more specific to individual locations, such as lakes and coastlines. What is so important in the development of these resources and therefore the economic success of the individual village is the cohesiveness of its community. In order to clear land for agriculture from the wilds, and then to keep the land productive and sustainably so, members of village communities in the past had to be supportive of one another. Whereas some of the earliest rural settlements in Britain would have been farmsteads and hamlets with inhabitants linked by kinship, as the population and settlement size both grew the community would have been made up of more than one extended family. The cohesiveness of the community and the ability of its members to work together probably passed through three phases. First came the **communality of agreement** whereby families agreed to cooperate for the common good, then followed the **communality of effort** that involved the practice of the agreement and the division of labour of the daily tasks within the community, and finally, by the time a village had developed into a large and more sophisticated unit, there was a **communality of enforcement** that would involve the formalisation of these practices in laws that might be imposed from outside and indeed be connected to some form of taxation.

The successful use of the intrinsic and extrinsic site qualities can be seen all over Britain, but one of the best illustrations can be found in the limestone and chalk hills of southern England, e.g. the North and South Downs, the Chilterns, the Mendips and the Cotswolds. In these areas **strip parishes** are a common phenomenon (see Figure 2.3). The territories attached to the villages stretch from hilltop to valley bottom in order to include within them all the different types of land available within the local environment. The exposed hilltop, often a plateau or gentle dipslope, was ideal for the rough grazing of sheep as it was suitable for little else. The steep hillside above the village was often left in a natural or semi-natural state as it was too steep for most forms of agriculture; however, areas of uncleared woodland provided fuel, building material, and somewhere for pigs to forage for acorns and other food. The village itself was located on the gentle footslope

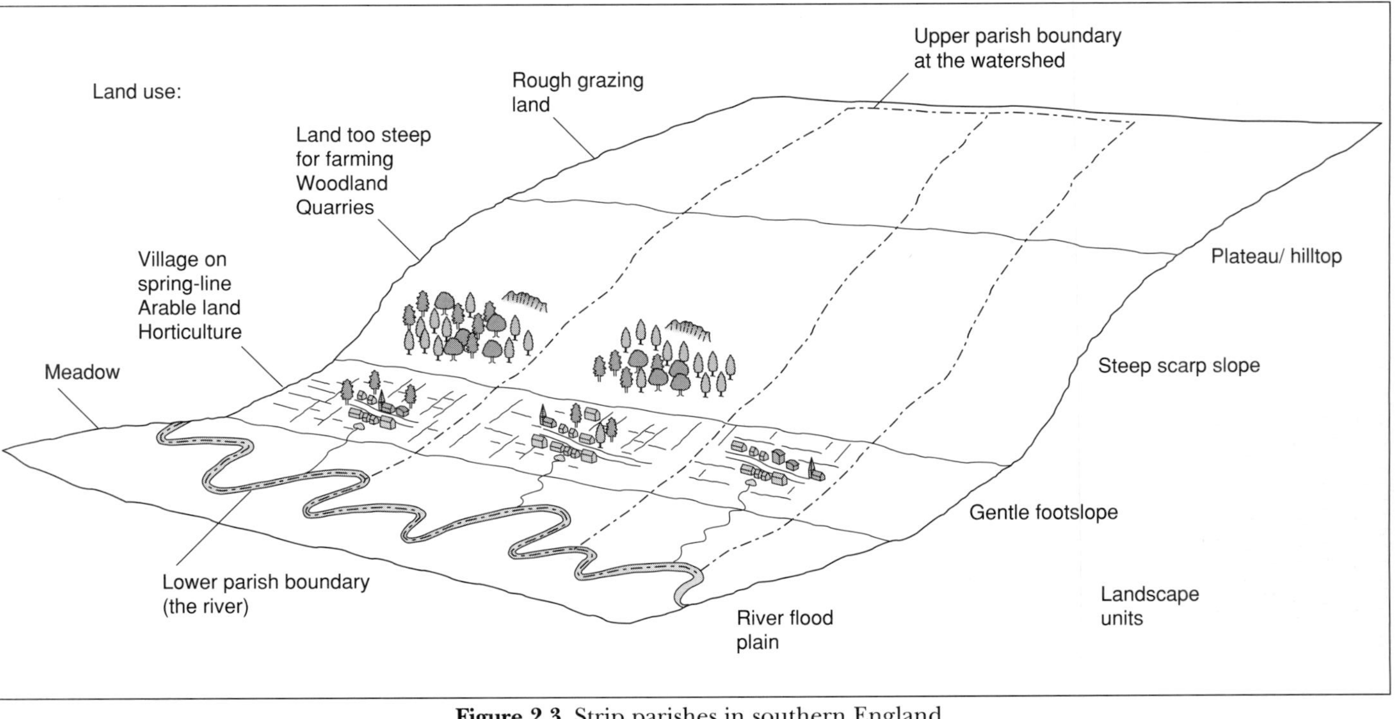

Figure 2.3 Strip parishes in southern England

beneath the hills; this had the advantages of a spring-line water supply, good drainage and shelter from strong winds. As most footslopes are made from colluvial material washed down by erosion from the hills, they provide richest soils on which the arable crops of the village can be grown, in close proximity to the village itself. The heavy clay soils of the valleys below and the abundance of water from the main river made them ideal meadow country for the grazing of cattle.

4 The Influence of Altitude Upon Settlement Location

Altitude has a major influence upon the siting of rural settlement. In Britain height above sea level imposes a major restriction upon settlements. Located between 50° and 60°N, shelter from exposure to cold winds from the north and east, as well as from exposure to low temperatures due to the altitudinal lapse rate, are prime considerations in settlement location. Few large nucleated villages are to be found at altitudes above 250–300 metres, and the figures are comparably lower in northern Scotland than in southern England.

In tropical locations, high-altitude settlements may be more favourably positioned than places at lower altitudes. Andean states, such as Peru and Ecuador, have their densest rural populations at high altitudes. Their lowlands both along their arid and semi-arid coastlands and their thickly forested eastern interiors are less attractive for agricultural developments and therefore rural settlement than the intermontane basins of the high Andes, which have both rich soils and temperatures moderated by altitude. In the Andes, therefore, it is not uncommon to find nucleated villages at altitudes between 3000 and 3500 metres above sea level.

In sub-tropical locations, such as the Mediterranean region, the situation lies somewhere between that of Britain and that of the tropics. A very interesting illustration of the significance of altitude in the location of settlements can be found in the southern Italian region of Basilicata. Table 2.1 shows the number of settlements located in each of 11 altitudinal zones. Although this table includes urban as well as rural settlements, as well over 95% of the places involved are rural centres, the inclusion of the towns does not alter the basic patterns observed. The most common type of rural settlement in the region is the *città contadina* (literally 'peasant city') – a large type of nucleated village perched upon a hilltop for reasons of defence and for protection from malaria, which until the 1950s was rife in the swamps in the valleys below and along the coastal plain. The only four settlements located below 100 metres are new service centres or holiday resorts created along the coast since 1950 and the eradication of malaria. The draining of the marshes and the construction of the new settlements were carried out through the

Table 2.1 The relationship between altitude and settlement location in Basilicata, Italy

Altitude	No. of Settlements
0–100 m asl	4
100–200 m asl	1
200–300 m asl	3
300–400 m asl	10
400–500 m asl	20
500–600 m asl	20
600–700 m asl	23
700–800 m asl	20
800–900 m asl	15
900–1000 m asl	11
over 1000 m asl	3

fund for the regional development of southern Italy, the '*Cassa per il Mezzogiorno*'. From Table 2.1 it can be seen that the vast majority of settlements are at altitudes between 401 and 800 metres, and that the optimum altitudinal band is between 601 and 700 metres. These high locations were colonised mainly between Roman and Mediaeval times when different waves of settlers moved from the mountains further north or inland from the coast – the latter included groups of refugees from what are today Greece and Albania.

Villages below 400 metres would have been much more prone to malaria than those above. Nevertheless, some of them still suffered; the two villages of Aliano (489 metres) and Grassano (515 metres), whose harsh way life was described vividly in 1935 by Carlo Levi in his autobiographical commentary, *Christ Stopped at Eboli*, both had a large proportion of their populations with malaria because they worked in the fields at lower altitudes. Even the children were not spared from the scourge of malaria and Levi described the situation thus:

> The children were pale and thin with big, sad black eyes, waxen faces, and swollen stomachs drawn tight like drums above their thin, crooked legs. Malaria, which spared no one in these parts, had already made its way into their underfed rickety bodies.

The spatial location of the rural settlements of southern Basilicata will be examined in more detail later on in this chapter.

CASE STUDY: THE ROLE OF *FENG SHUI* IN THE LOCATION OF CHINESE RURAL SETTLEMENT

In China *feng shui* principles have long been used to probe the landscape, look at the regularities and asymmetries within it, and then decide upon a suitable location for settlement. The same principles were used for humble farmsteads and villages, right up the hierarchical ladder to the great planned imperial capitals, such as the Forbidden City in Beijing. Even today some new high rise office blocks in the booming port city of Shanghai have *feng shui* principles applied to their sites before construction. *Feng shui* literally means 'wind and water' and concerns people living in harmony with nature. The concept, often translated in English as **geomancy**, has been practised in China for at least 3000 years. Books upon the principles of farm and village location go back to the third century BC. In Europe, by contrast, virtually nothing is known about what ritualistic principles lay behind the locating of settlements, other than that recorded by Roman writers such as Vitruvius.

Although so many of the principles of *feng shui* are now regarded as being based upon little more than superstitions of the past, large numbers of Chinese villages have been built in what are both highly practical as well as beautiful locations. Writings on the *feng shui* mention such concepts as *cangfeng*, which represents shelter from the north-west winds (which would be blowing from the snow-clad mountains of the Asian interior), and *deshui* or a location near water.

As in other Asian cultures the importance of a balance within nature between opposites is taken into account with village location; thus *kanyu* or the linkage between heaven and earth is an important locational consideration. The relief pattern of an area was also of great significance when locating a village or farm site. A series of hills resembling mythical creatures such as dragons were regarded as auspicious as were meandering rivers. These locations not only had the site advantages of shelter from the winds provided by the mountains and rich soils provided by the meandering river, but also had the great aesthetic beauty seen in many traditional Chinese paintings. Figure 2.4 illustrates this type of location well. Creating a serene and comfortable environment in which to live was always a very important aspect of the Chinese tradition and philosophy. Also, the practice of *feng shui* principles has probably been very been good for the environment. As Knapp (1992) states:

> ...*feng shui* no doubt has helped restrain Chinese villagers from unwise ecological decisions, nurturing reasonably sound ecological practices and leading to 'planned' settlements far ahead of their time.

(a) Located according to traditional 'feng shui' principles

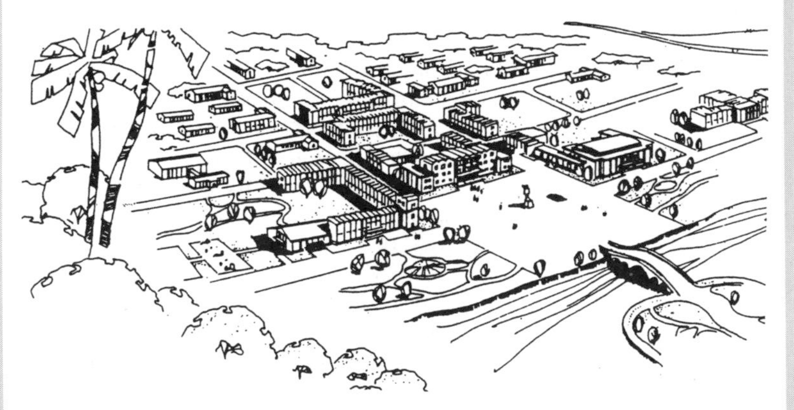

(b) Modern communal village (1970s)

Figure 2.4 Chinese village location in relation to the landscape (after Knapp, 1992)

Despite the upheavals created by the Cultural Revolution in the late 1960s and early 1970s, and the foundation of thousands of new, soulless communal villages resembling high rise industrial suburbs, village locations in much of rural China still retain their *feng shui* based origins. Furthermore, elements of the *feng shui* code of practice remain reference points for the Chinese villagers for whom they provide a form of mental map.

CASE STUDY: RURAL SETTLEMENT LOCATION IN THE MENDIPS

The Mendips are a range of Carboniferous Limestone hills located in the northern part of the old county of Somerset, and provide a very clear example of the way in which physical factors influence both the site and size of rural settlements. (Figure 2.5 shows the location of the main nucleated settlements in the Mendip area). The Mendips are formed of a limestone anticline that has steep slopes to both the north and the south. These slopes are most marked in the west and central parts of the Mendips but the hills merge into more gentle, rolling topography in the east towards the town of Frome. In the central part of the Mendips the hills extend up to a broad plateau top, where they reach their highest point, Beacon Batch (325 metres). On the plateau the underlying strata of Old Red Sandstone are exposed and this has given rise to an area of bleak, windswept, bracken and heather covered upland. Most of the rest of the top of Mendip was converted to farmland in the period of parliamentary enclosure (1750–1850).

The most favourable places for settlement are on the gentle footslopes of the hills, where there are both spring lines and rich colluvial soils formed from deposited materials that have been washed downslope over thousands of years. In some places, these footslopes are much more extensive than in others. The dominant feature of the pattern of village location in the Mendips is the two strings of spring-line villages to the north and south of the hills, at the foot of their steep slopes.

The southern spring-line settlements tend to be bigger than those to the north as they have the advantages of aspect (on a south-facing slope), as well as generally having a larger area of fertile land on the footslope. Of the places located on the southern spring line, Wells developed into a market town and cathedral city, and Shepton Mallet and Axbridge were granted *burgh* or market town status by the Middle Ages, although the latter is now just a village. Cheddar, the other large settlement on the southern footslope, although still a village, has the population size of a small town. These southern villages open out onto the Somerset Levels, which were liable to winter flooding particularly up until the 1750s when large-scale reclamation projects were carried out. The villages are therefore located at altitudes of 10–50 metres, afforded by the colluvial footslopes.

Some of the villages on the north-facing slopes are considerably higher up than those of the south-facing slope (e.g. Shipham is located at 125 metres and Blagdon at 140 metres); this leaves them potentially exposed to cold northerly winds, which helps to

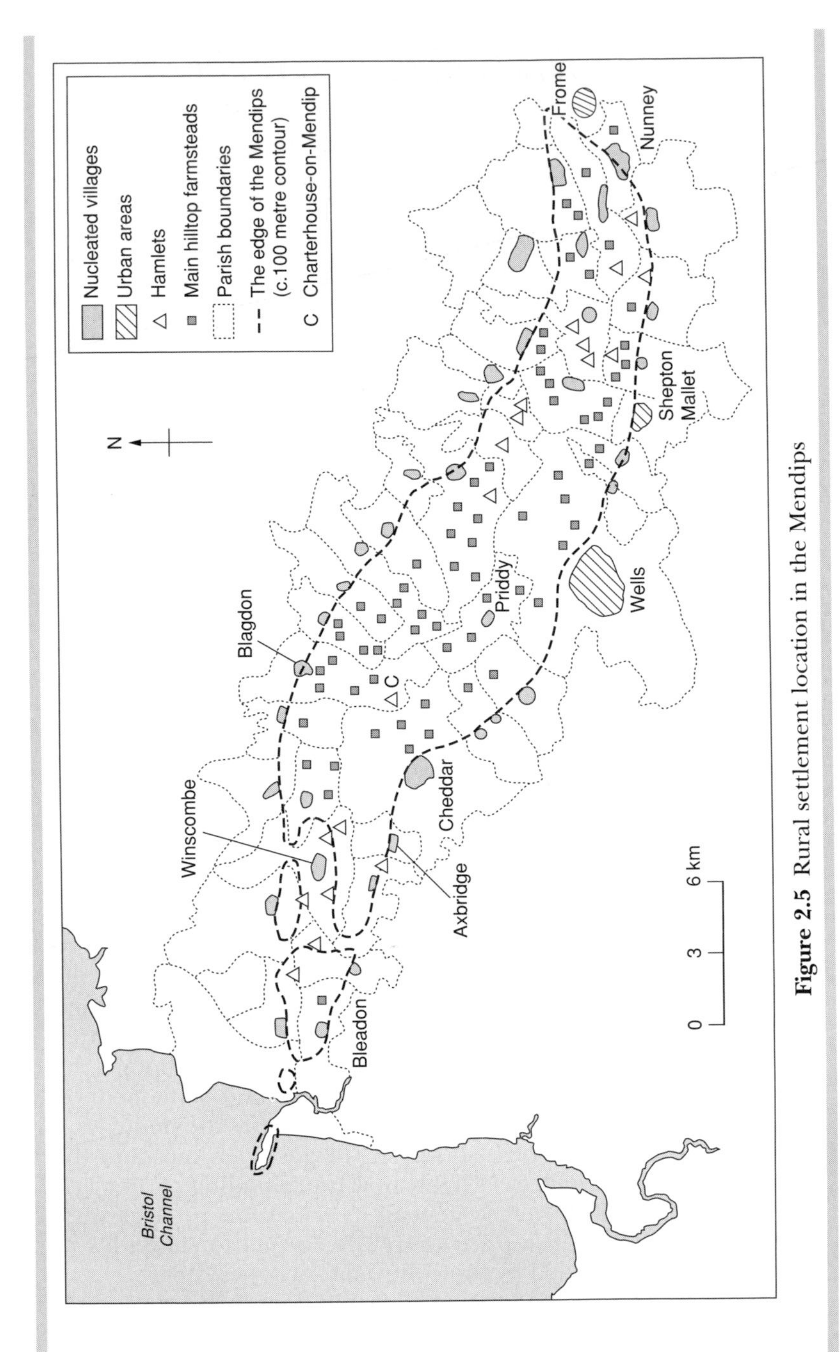

Figure 2.5 Rural settlement location in the Mendips

explain why they were not so successful in terms of physical growth and why they failed to develop into market towns in the Middle Ages. Until a few decades ago all of the northern footslope villages were much smaller than they are today. Some of them, such as Winscombe, have expanded very rapidly because of their favourable position on the main commuter routes to Bristol and Bath.

Only two nucleated settlements are located on the top of the Mendips: Priddy and Charterhouse. These can be classified either as large hamlets with churches or small villages; both are located in shallow dry valleys at an elevation of around 250 metres. Charterhouse was a mining settlement in Roman times because of its valuable lead deposits, it then became a monastic farm in the Middle Ages and lead mining was revived in the nineteenth century. Today the settlement is just a small group of farms. Priddy evolved as an important sheep-farming and marketing village and its community survives today as a cluster of farms, houses, two pubs and a limited number of other services, located close to its village green. Priddy benefits from its location on the limestone plateau and on long-distance footpaths as a centre for potholing, rambling and other leisure pursuits.

Dispersed farmsteads are found all over the Mendip region. Before the eighteenth century there were few on the more exposed parts of the Mendip plateau; those that existed were attached to old enclosures in slightly sheltered hollows and dry valleys. With the enclosure of the hilltops in the eighteenth and nineteenth centuries, a whole network of dozens of new farmhouses were established, from which the newly enclosed land was worked. The farms in the highest locations are found between the 275 metres and the 290 metres contour in the hilltop parish of Priddy (e.g. Hill Farm, Spring Farm and Swallet Farm). Dispersed farmsteads are also found around each individual village; these either date from very early times or from the period of enclosure when the open common arable fields were split up and divided between the new farms. A third location of dispersed farmsteads is on the outliers of the Mendips: the small hills that lie on the Somerset levels to the south of the main anticline, (e.g. Lodge Hill and Nyland Hill). Here farms are located around the edges of the outliers on slightly elevated land (10–20 metres) above the normal winter flood levels (e.g. Quarry Farm, Decoy Pool Farm and Rookery Farm located at the foot of Nyland Hill).

CASE STUDY: RURAL SETTLEMENT LOCATION IN SOUTHERN BASILICATA, ITALY

The southern Italian region of Basilicata has been looked at earlier in this chapter as an example of how altitude can influence the location of rural settlements; here it will be considered in the more general context of what factors help to determine the sites and positions of Mediterranean villages.

Figure 2.6 shows both the physical setting and the distribution of settlement in the southern part of Basilicata, close to the Ionian Sea. The area is very different from that of the Mendips. Along the Ionian shore there is a narrow coastal plain, which stretches just over 5 km inland at its widest part; this area has only been colonised since the eradication of malaria in the 1950s. The rest of the area is dominated by the hills of young sedimentary rocks and the valleys of the five rivers that cut through them. The hills, that rise up to almost 900 metres in several places, are made of incoherent strata of clays, mudstones, shales and sandstones, these are heavily faulted and subject to slope failure and gulley erosion. One of the villages, Craco, had to be abandoned in 1975 and relocated in the valley below because of its damaging landslides, and others, such as Pisticci, have lost several rows of houses over the edges of the hilltops upon which they are located because of slope failure. The river valleys of the Bradano,

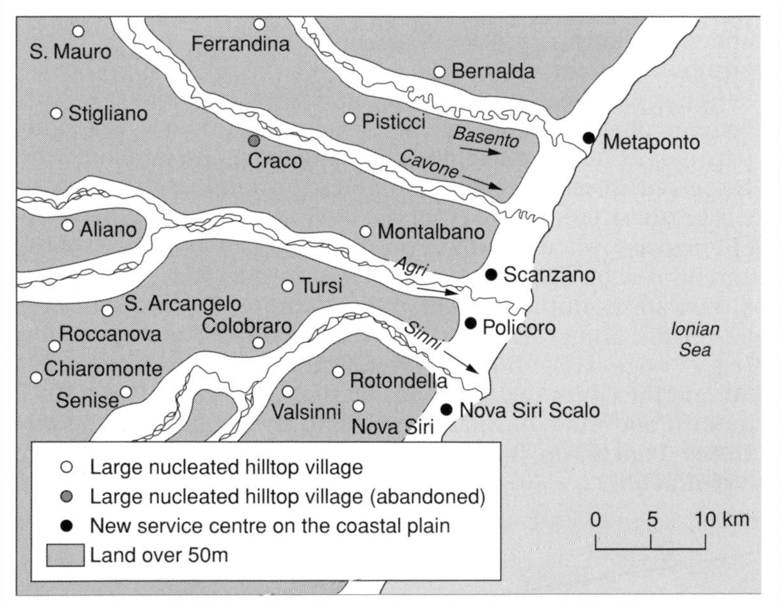

Figure 2.6 Rural settlement location in south Basilicata

Basento, Cavone, Agri and Sinni are 2–3 km wide and have steeply rising valley slopes. The rivers themselves are known as *fiumare,* which in Italian signifies a stream that is highly variable in its seasonal discharge. In the summer the rivers are almost dry, and in the winter they from time to time go into spate following heavy seasonal rainfall. The valley floors are strewn with deposited gravels brought down from the mountains, which have resulted in the rivers having heavily braided channels. Until the 1950s these valleys were highly malarial and devoid of settlement.

Although the coastal plain had ancient Greek settlements along it, the heavy deforestation in the mountains and subsequent soil erosion and development of marshes along the coast led to the abandonment of these sites. In the Mediaeval period the population sought refuge in the higher altitude locations of the hills, where it was less vulnerable to both malaria and the attacks from Saracens and other invaders. The main older rural settlements are the large nucleated villages perched along the ridges of the hills between the main river valleys where well over 90% of the local population lived. Until recently they were just linked by tortuous roads that made journey times between them slow. Between these villages are large isolated estated farms or *masserie* and a few hamlets, some of them based upon small monasteries. Many of the large nucleated villages have seen heavy depopulation ever since the late nineteenth century.

The fortunes of the region changed as a result of the work of the *Cassa per il Mezzogiorno,* Italy's regional development fund for the south, which operated between 1950 and 1985, and has subsequently been replaced by a succession of agencies dealing with the 'Southern Question'. The changes brought about by the regional development programme have greatly added to the settlement pattern, as well as changing the local infrastructure. The eradication of malaria, the building of new roads, land reform, the introduction of new irrigation schemes, the flood control of the main rivers, the development of the local natural gas deposits and the tourist potential of the coastline have all led to a transformation of the area.

The new elements in the settlement pattern of southern Basilicata are:

- hundreds of new farms attached to smallholdings that produce high yielding crops, on reclaimed land along both the coastal plain and the valley bottoms
- new service centres, like large villages, built along the coastal plain to act as market and agricultural processing centres for the farms (e.g. Scanzano, Nova Siri Scalo)
- new small-scale seaside resorts along the sandy coastline of the Ionian Sea (e.g. Lido di Scanzano, Metaponto)
- new industrial settlements in the valleys close to the gas deposits (e.g. Bernalda Scalo)

5 The Changing Location of Rural Settlement

The concepts of continuity and change were dealt with in some detail in the previous chapter. What is clear about settlement location is what may be suitable for one group of people may not be successful for another. In many pre-historic European and Asian societies it was common practice for a group of people to move their settlement to another place when a member of the community died, especially in the case of the head of the family or clan. This mark of respect, or reaction based on fear of the spirit world, would have been much easier when populations were much sparser, and rural settlements were small and more mobile.

Since the sedentarisation of most rural settlements in most parts of the world, the changes in location of a settlement have been more associated with upheavals that have resulted from either natural hazards or human conflicts. Many of these factors are dealt with in more detail in Chapter 5 in relation to settlement abandonment, but some of the main causes of the relocation of rural settlements are as follows:

- climatic change influencing the upper or lower altitudinal limits of agriculture and settlement
- the perception of hazards such as flooding causing a community to relocate in a safer site
- a natural disaster such as an earthquake necessitating the move to another site
- the creation of more fertile farmland by the clearing of forests or through land reclamation encouraging a move to a more productive location
- new technology allowing a society to develop new agricultural resources and settlements
- population growth and the subdivision of land holdings forcing part of a community to move to new territories and establish new settlements
- tribal or civil warfare or invasion from outside encouraging a move to a more readily defensive site
- a calculated move instigated by a landowner for either political or economic reasons
- government action replacing old agricultural settlements with new 'model' ones for ideological reasons.

CASE STUDY: THE CHANGING LOCATION OF RURAL SETTLEMENT ON THE BANDIAGARA ESCARPMENT, MALI

In the Sahelian state of Mali on the southern fringe of the Sahara, live the Dogon people, who are one of Africa's most traditional rural societies, little altered by modern influences from outside. The Dogon villages consist of clusters of mud-built family compounds, with rooms and granaries centred around courtyards. In the middle of these villages are located various buildings of communal importance such as the *ginna* or fetish house and the *toguna* or elders' meeting house. There is a string of a dozen Dogon villages including Yayé, Tirelli, Ourou and Dourou, which stretches for some 50 kilometres along the foot of the Bandiagara escarpment, that have a north-west-facing aspect. The actual sites of these Dogon villages are on a fairly steep slope of up to a 45°, scattered with debris from different types of mass movements and including huge boulders from rockfalls; how these fit into the Dogon perception of hazards is rather a mystery. Below the villages on the plains are the Dogon millet fields and the wells that are their main sources of water. Animals such as sheep and goats are normally kept within the villages but are also taken to graze on the millet stubble after the harvests.

Although the Dogon villages have been in their present location for about 500 years, this represents a shift from the earlier settlements that existed upon the Bandiagara escarpment. The Dogon are directly descended from the Tellem people, the remains of whose settlements are still visible today. The ruins of the mud-built circular tower houses of the Tellem are located in naturally occurring niches high up in the limestone cliff face of the Bandiagara. Today these niches are used as burial sites for the Dogon; the dead are therefore ritually assigned to the places of the ancestors. Virtually all of the present-day villages are backed by Tellem settlement ruins and this would suggest that there was a deliberate shift in the location of rural settlements at some stage around 500 years ago. What exactly caused the movement to a lower location is difficult to deduce from any archaeological evidence or from the legends of the Dogon themselves. Various possibilities for this shift could include:

- the drying up of water supplies such as springs that may have existed higher up on the cliff faces
- the passing of the society from a turbulent period of inter-tribal warfare to a more peaceful and stable one
- a rapid expansion of population that led to the need of much greater living space.

Questions

1. What are the main factors that influence the distribution and location of rural settlements?
2. Why are some rural settlement locations likely to be more successful than others?
3. Using specific examples and case studies, explain why the location of rural settlements can vary considerably from one region to another.
4. Contrast the differences between *intrinsic* (site) and *extrinsic* (position) factors in the determining of rural settlement locations
5. Why may the locations of rural settlement change through time?

Summary Diagram

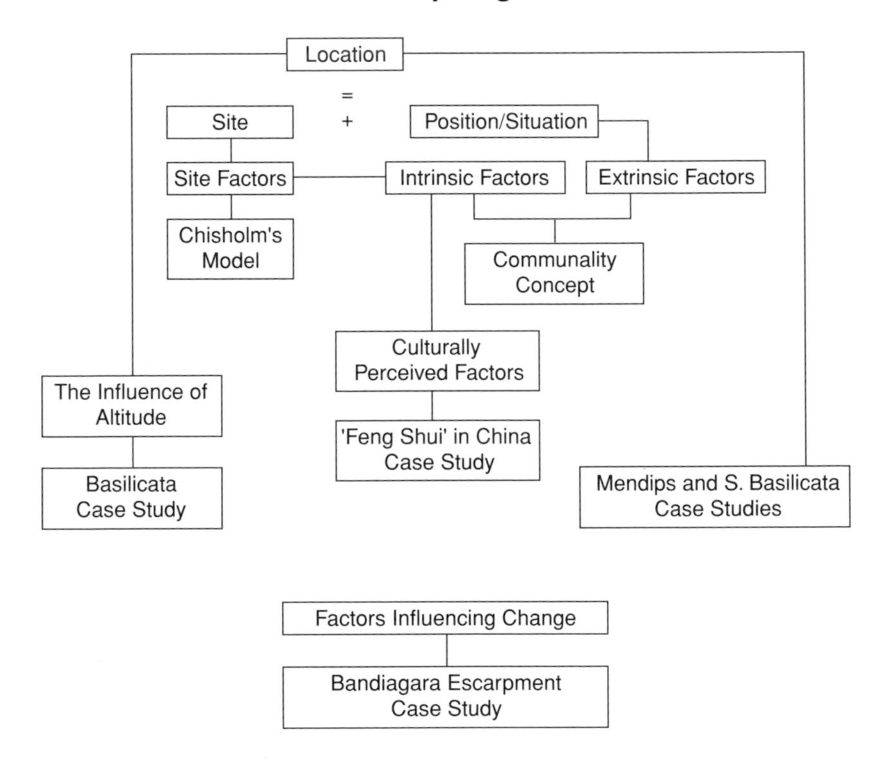

3 Rural Settlement Form

KEY WORDS

Morphology: the internal layout, plan or structure (of a settlement).
Agglomerated: clustered, grouped together.
Linear settlement: a village following the line of a road, river, valley or causeway.
Settlement accretion: the growth of a settlement by the gradual adding on of new housing.
Settlement infill: the growth of a settlement by the addition of new housing within its internal open spaces.

A man's village is his peace of mind

Anwar al-Sadat

1 Introduction

The internal layout or **morphology** of a rural settlement is determined by two main groups of forces: those imposed by the natural environment within which they are located and those resulting from the degree of organisation imposed by the human inhabitants, past and present.

Natural landscapes present both opportunities and restrictions to rural settlement layouts, in much the same way as they influence location. Flat or gently sloping land that is not regularly flooded, such as a river terrace or slightly elevated land close to the sea, provides an ideal landscape in which a village or hamlet can be laid out in any way desired by the founding inhabitants. By contrast, mountains, deep valleys, extensive areas of marshland and rugged rocky coastlines may impose heavy restrictions on the initial layout and then the subsequent development of rural settlement morphology.

The human impact upon rural settlement layout is as diverse as the number of different economies, societies and cultures to be found throughout the world. Some of the main human factors involved are:

- land tenure, land ownership and the role of the chief landowner
- the role of religion, ritual, superstition and tradition in the foundation of settlements
- the defensive requirements of a settlement
- the nature of the farming type or other form of economic activities taking place within the settlement.

2 The Classification of Rural Settlement Types by Form

It is virtually impossible to make a detailed classification of rural settlements to suit all parts of the world. Cultural variations have given each country or region its distinctive settlement forms. Some of these forms are repeated in many different parts of the world, one of the best examples being the **linear village**, which is generally the product of either a restricted physical site such as a narrow valley bottom or the development of houses along a routeway. Classifications have tended to be made by individual geographers to fit their own country. One of the first attempts to put rural settlement form into a world perspective is found in Vidal de la Blache's *Principles of Human Geography* published in 1926. The work considers rural settlement types in northern Europe, the Mediterranean region, China, Indochina and India; in the end de la Blache concludes that 'These examples demonstrate the fact that the type of distribution is determined by the region itself'. Rather than relying on any other generalisations, therefore, it is better to examine a number of different examples both within Europe and beyond.

3 Rural Settlement Types in Germany

There are considerable variations in rural settlement forms in Germany, resulting from the different land qualities of the country, the successive waves of colonisers settling the land at different stages in history and demographic changes, particularly population growth.

The earliest agricultural settlements still surviving are the small-scale nucleations established on the lighter soils in parts of northern Germany. The term used for this type of hamlet, which possibly dated back to Celtic and Roman times, is *Drubble*; closely associated with each *Drubble* were small rectangular fields irregularly rotated known as *Blockflur*. At some stage these became replaced by a single arable field known as an *Esch*, which was subdivided into individual strips worked by different members of the community. By around the eleventh century the population grew greatly, new lands were colonised and the older forms of settlement became widely replaced by much larger nucleated villages in many parts of Germany where richer soils permitted. The term given to this type of rural settlement in German is *Haufendorf*. These much larger settlements required much greater food production and were therefore associated with field systems known as *Gewannflur*; these were large open common fields divided into strips, similar to those in Britain from Saxon times onwards.

Another form of German village that dates back at least 1000 years is the *Terp*. This type of village is small and roughly circular or spiral in form and was built on top of artificial mounds constructed within coastal marshes. Similar villages are found over the border along the Dutch coast. As colonisation continued in the western parts of Germany following the rapid population growth between the tenth and the thirteenth centuries, more of the feudal lords established well-planned villages in areas that were still in their natural state. Two of the most common types of regular linear village established in this way were the *Waldhufendorf*, or forest village found particularly in the Black Forest, and the *Marschhufendorf*, or marsh village found along the North Sea coastal areas that had been reclaimed by the building of dykes.

The eastern parts of Germany were colonised later than in the west. Here there was a much sparser, mainly Slavic, population whose lands the Germans were conquering and also vast tracts of uncleared forest. In these eastern lands the Germans established a number of distinctive planned village forms in the period between the twelfth and fifteenth centuries. In the lowlands linear villages were the most common; these included the *Angerdorf*, an elongated village with a central communal green, and the *Strassendorf* or roadside village. In addition to these the *Waldhufendorf* was widely established in the wooded hills and uplands. Less rigidly planned but still common on the lowlands of eastern Germany, and particularly located at sheltered points upon a break in slope, is a round village form known as a *Rundling*. This type of small village or hamlet has a series of farmhouses arranged around a central green that could have acted as a

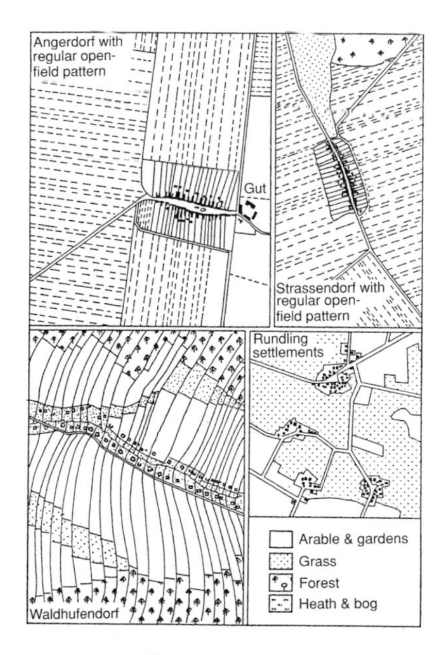

Figure 3.1 Characteristic German village types

defensive corral for livestock. It is uncertain whether this settlement type is of German or earlier Slavic origin. New rural settlement forms were still being imposed upon the German landscape during the seventeenth and eighteenth centuries. Pressure upon land was resulting in the further colonisation of marginal lands and the reclamation of marshes. In the age of Frederick the Great, there were large new villages established in the marshlands of Prussia, as well as in north-west Germany; these were known as *Fehnkolonie* and were made of a series of linear forms at right angles to one another. Figure 3.1 shows some of the most characteristic German rural settlement forms.

4 Rural Settlement Types in Britain

One of the first works to go into a detailed study of British villages was *The Anatomy of the Village* by Sharp in 1946. This was written at a time when the motor car was starting to have an impact upon rural Britain and also there was considerable demand for new housing in villages due to a rising population and social changes (e.g. nuclear families were becoming more common than extended ones). Sharp studied the layout of numerous villages in England, Wales and Scotland, and in doing so came up with a very simple classification of village form. He recognised four basic village types:

- The roadside village, where its form was linear due to its construction along a routeway, e.g. Shincliffe, Co. Durham and West Wycombe, Bucks.
- The squared village, which actually included many different non-linear forms, such as the the hollow square, oblong and triangular village, with or without central greens, as well as infilled square, e.g. Heighington, Co. Durham (hollow square), Coneysthorpe, Yorkshire (oblong), Writtle, Essex (triangular) and Sherston, Wilts (infilled square).
- The planned village, where at some time in the last few hundred years a landowner has built a totally new village to accommodate his tenants and in doing so created a model village with good-quality housing, e.g. Milton Abbas, Dorset (1786), Harewood, Yorkshire (1760) and Lowther, Cumbria (1682).
- The seaside village, which was often very irregular in form because of the configuration of the coastline, e.g. Polperro, Cornwall. W.G. Hoskins, in his inspirational work *The Making of the English Landscape* carries our understanding of village morphology further. He states:

> The variety of plan among the villages of England...is profoundly interesting – and tantalising – to the historian of the landscape. It is interesting because he realises that this variety of forms also certainly reflects very early cultural or historical differences.

> ... we cannot be sure that the present plan of a village is not the result of successive changes which were completed before the earliest maps are available: we cannot be sure what the *original* shape was in many instances.

Hoskins observes that by the tenth century most English villages had been established as permanent settlements in their present locations and, where it has not been obscured by twentieth century developments, their original fundamental layout is still retained. The first detailed maps of villages were drawn up around the late sixteenth and early seventeenth centuries, and these provide evidence of the continuity of village layout over the last few hundred years, although this is not necessarily an indication of what the original shape was in Anglo-Saxon or pre-Saxon times.

Hoskins recognised three major categories of villages in England:

- the village grouped around a central green or square (e.g. Finchingfield, Essex and Easington, Co. Durham)
- the village strung out along a single street (e.g. Long Melford, Suffolk and Henley-in-Arden, Warwickshire)
- the village that consists of dwellings planted down haphazardly, with no evident relationship to each other (e.g. Middle Barton, Oxfordshire).

Hoskins also gives an important insight into the origins and the purpose of the village green:

> It seems likely that these villages built around the perimeter of a large green or a square represent enclosures for defensive purposes, like the native villages of some East African tribes today. Here the huts are grouped round the perimeter of a circular pound, with narrow openings between them, which are closed at night by thorn fences ... Into these pounds the livestock are driven at night. In the villages of Saxon England, the necessity for the protection from wolves may have led for the same plan being adopted.

Roberts (1987, 1996), has produced the most detailed classification for British rural settlements to date. His taxonomy (Figure 3.2) is based upon three groups of principles: the **basic shape**, the **degree of regularity** and the presence or absence of a **village green**. The basic shape is classified as being **linear**, either in the form of a street village or based upon houses arranged in rows or **agglomerated** either on a grid-like pattern or on a circular or radial pattern. The degree of regularity of a village depends upon two things: to what degree the settlement was consciously given a planned layout when it was founded and the extent any regular plan that existed in the past has become obscured by subsequent developments. This is why much older settlements are likely to appear to be irregular in comparison to villages of more recent foundation. Although the villages established by landowners in the seventeenth to the nineteenth centuries display

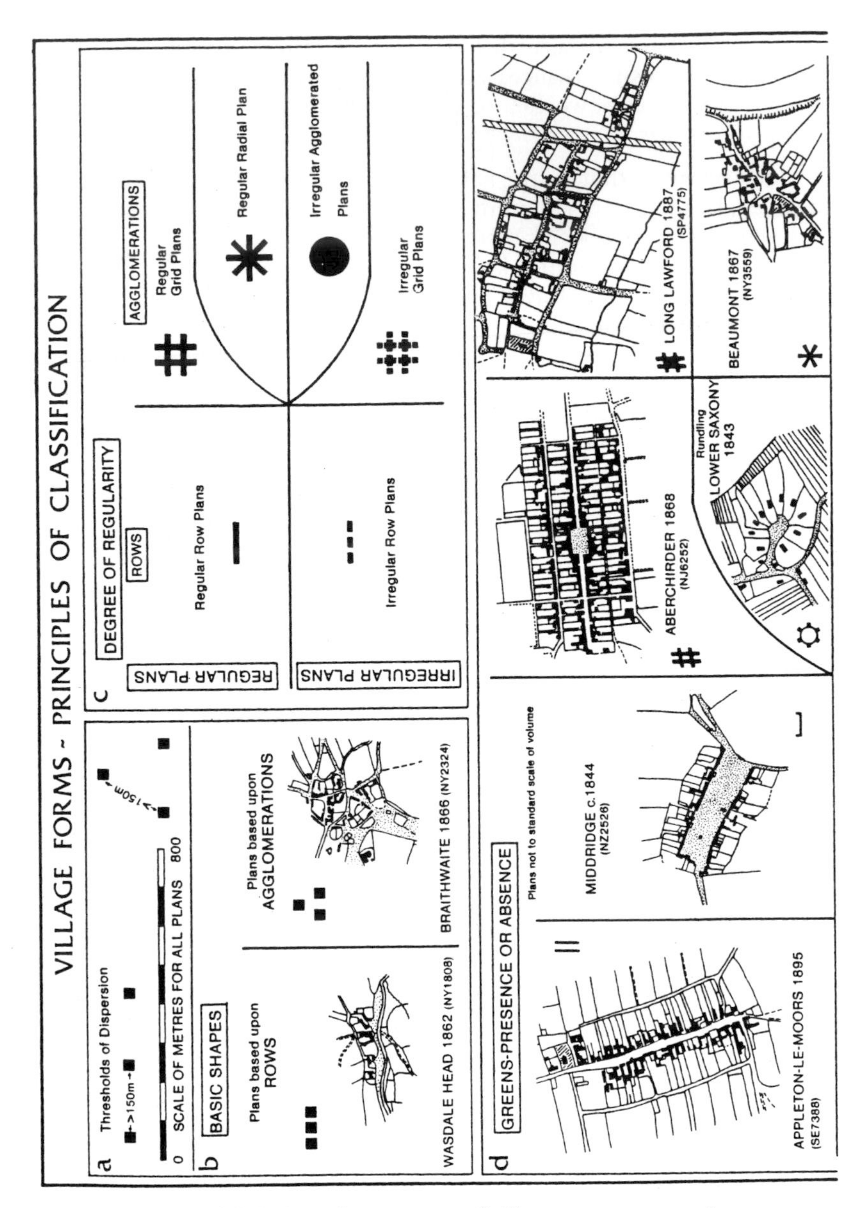

Figure 3.2 Roberts' taxonomy of village types contrasting British village layouts (after Roberts, 1987)

the greatest degree of regularity in British village layout, some going back 1000 years have a high degree of regularity. Many of the 'toft' and 'croft' (garden plot with a single house built on it) Saxon or Viking settlements arranged around a village green show considerable regularity of plan. This reflects the communality of early settlements,

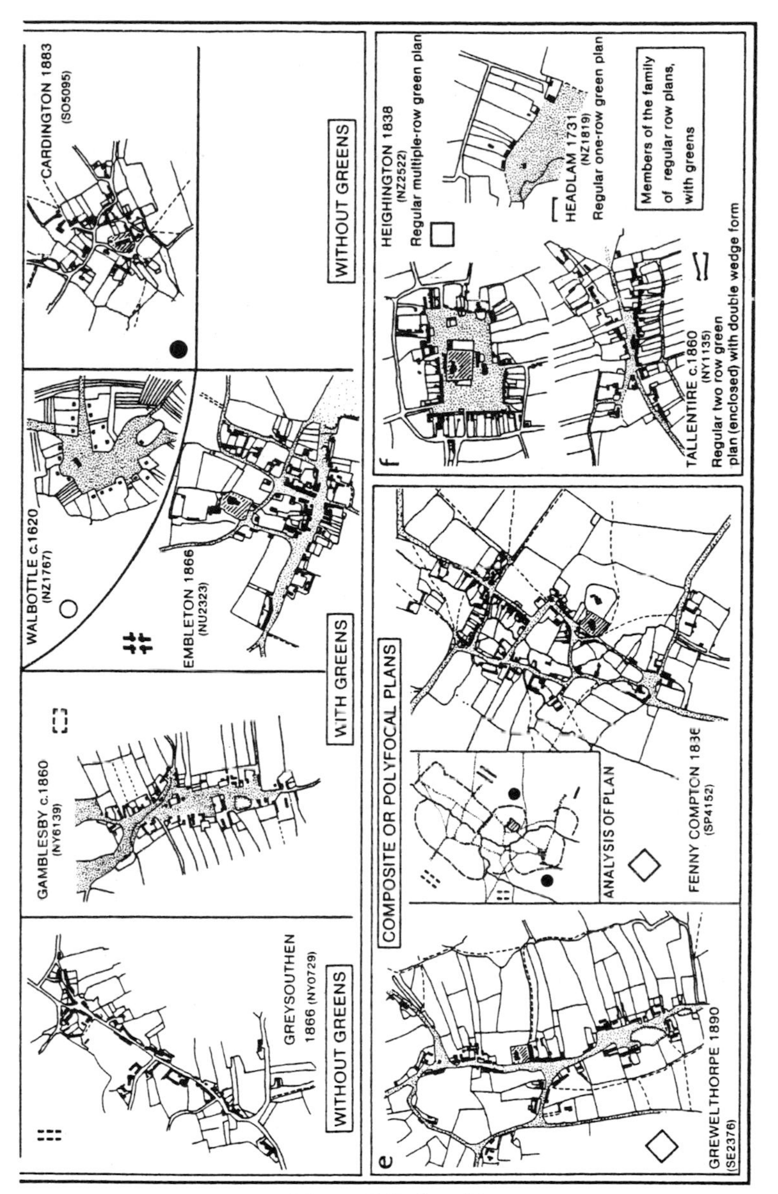

Figure 3.2 Cont'd

families working together and the importance of each family having its 'fair share' of the land available, whether it was owned communally or by a single landowner. To what extent any ritual practices were involved in the layout of settlements has been obscured by time and the lack of written records. There is evidence from Scandinavia that when open common fields were being divided up into strips this was done by using *solskifte* (sun-division), where the passage of the sun overhead was used as the basis for the orientation of the subdivision of the larger fields. It is possible that similar natural phenomena were used in the laying out of village plans, but there is no evidence for this.

The presence of a village green is generally the reflection of the type of economy practised within a particular settlement. Those settlements in which a pastoral economy was particularly important, where security was a potential problem or indeed where a communal central space was needed for special functions would have found greens of great value. Figure 3.2 shows some examples of differing British village plans.

When examining village plans three points that Roberts mentions are important to bear in mind:

- plans reflect the relationships between three types of space: **private space**, i.e. that taken up by individual houses and gardens, **public space**, e.g. the roads and lanes, and the **communal space** that would include the village green and water supplies; some areas such as the church would come into all three of these categories
- plans do not necessarily reflect **building density**, which may sometimes make it difficult to compare one village with another in terms of its population and therefore importance, e.g. the number of storeys a house has is not visible on plans
- plans change and evolve through time as a settlement gains or loses population and this adds to the complexity of layouts (this will be considered in more detail later on in the chapter).

CASE STUDY: EUROPEAN COLONISATION AND SETTLEMENT TYPES IN THE USA

The range of rural settlement types in the USA reflects the various layers of migrations that have superimposed their cultures upon the landscape through time. In the SW the legacy of the pueblo Indians adobe villages is still greatly in evidence, whether they are abandoned sites such as those at Mesa Verde, or still inhabited today as at Taos. From the sixteenth century onwards migrations of Spanish, French, English, Germans and Scandinavians added their cultural identities to the human landscape of North America.

It is at the level of the individual farmstead that different European styles are most recognisable. Areas settled by Germans and Scandinavians have a tradition of wooden farmhouses, whereas the Irish, Cornish and Breton settlers favoured stone-built farmhouses. Actual farm layouts, with specialist buildings and different styles of courtyards, also reflected the cultural identity of the migrant groups (e.g. the summer kitchens in German farmhouses and saunas in Finnish ones).

Grid pattern layouts are almost ubiquitous within rural settlements of the USA, because of the way in which the land was surveyed and staked out under the 'national land ordinance'. However, there are examples of changes to this basic rule such as the Welsh community in Illinois that deliberately resurveyed the land to produce irregular parcels of land reminiscent of those in Wales. In some German-settled areas of the Mid West the *Strassendorf* village form was transplanted from the Old World to the New. The most rigid and well-planned grid pattern villages in the USA are probably those established by strict religious groups – of which there are and were many, including the German Lutheran Amana in Iowa, the Swedish Jassonites in Illinois, and the Amish and Mennonites in Pennsylvania.

5 Rural Settlement Types in China

Knapp (1992) in his study of Chinese landscapes recognised ten different rural settlement regions within that vast country with its great variations in natural environment. For each of the regions he identified the relationships between village densities and village size. The two most extreme situations are to be found in the plains to the north of Beijing, where villages are large (with an average population of 1000) but well spaced (24 per 100 km^2) and the mountainous area of Guangxi in SW China where villages are small (average population *c.* 80), yet they are closer together (105 per km^2).

He also identified within China five basic types of rural settlement layout, although within each category there is considerable regional variation:

- compact villages (*tuanxhuang*), which are roughly round to polygonal in plan, are sometimes surrounded by walls and are the most common type of Chinese village found especially in low-lying areas
- nucleated villages (*jicun*), which are smaller than the compact villages, but have their houses closely packed together, tend to lack overall planning and green spaces
- linear villages (*daizhuang*), which are most commonly found along roads, rivers and canals, often built along artificial levées and can be as much as 5 km long
- ring villages (*huanzhuang*), which are not very common in much of China, but are to be found in locations where physical geography is favourable, e.g. within the meander of a river or upon a conical hill

- dispersed villages (*sanjun*), which are the equivalents of European hamlets, they are especially associated with mountain areas and may occur in both regular and irregular forms.

There is not space here to deal with all the regional variations in China, but two examples of the great diversity of rural settlement are worthy of mention. In northern central China along the course of the Huang He (Yellow) River there are thick loess deposits that have given rise to a whole series of unique settlement types in Gansu and Shaanxi provinces. In this region of temperature extremes the tradition has been to excavate houses from the soft loess deposits offering shelter from both the summer heat and the severity of the winter. There are a wide variety of ways in which houses are created and this influences the overall layout of the villages. Four main house types are recognisable in the region:

- pit cave dwellings, often invisible from ground level
- cliff cave dwellings set into the terraced hillsides
- earth sheltered dwellings built above ground but given barrel-vaulted roofs
- semi-below and semi-above ground dwellings built into the hillsides.

The other unique village type in China is located in the south-eastern Fujian province where from the third to the seventeenth centuries AD the Hakka people migrated and established their own style of villages. These are generally strung out along riversides and consist of a number of round and square fortress like houses. Each house is three storeys high and looks inwards into a central balconied courtyard with just one entrance to the outside world. Each housing unit has up to 150 people living in it and also has quarters for chicken and pigs. These villages represent one of the most traditional forms of communal dwelling still surviving today in China.

Chinese traditional villages were frequently the victims of the upheavals that took place during the period in which Chairman Mao was in power. Most changes occurred between 1958 and 1964 when peoples' communes were established and then from 1964 until 1978 during the 'Cultural Revolution'. Thousands of traditional villages were replaced by new Communist-style settlements and thousands of new settlements were also built from scratch. The villages of the 1960s and 1970s have the appearance of urban settlements with terraces of two- to four-storey blocks, laid out in regimented barrack-like rows and communal buildings such as assembly halls, theatres, schools and libraries that resemble industrial units.

6 Rural Settlement Layout in Indonesia

It is through the study of highly organised rural settlements in certain LEDCs that we can best understand the types of roles played by religion, magic and superstition in the laying out and planning of villages millennia ago in Europe. Our knowledge of the old religions in Europe has faded away but anthropological studies in countries such as Indonesia have revealed the importance of the relationship between religion and village planning.

Indonesia, which stretches for over 5000 km from east to west and is made up from over 13 000 islands, has perhaps the greatest cultural diversity of any country in the world. Despite this diversity there is are some similarities in the ordering of villages on islands that are far apart. What is important in many parts of Indonesia is the polarity between mountains that represent the divine and either the sea or low-lying plains that represent the demonic. At the same time villages are often divided into two rows of dwellings that face each other; these represent pairs of opposites such as sun and moon, male and female, upstream and downstream, gold and silver – themes which are found in cosmology throughout South-east Asia.

Figure 3.3 shows the simplified layouts of several types of Indonesian villages. The duality of the cosmos is seen clearly in the traditional Aga villages on the island of Bali, where the eastern row of houses is associated with the temple of origin or life and the western row is connected to the temple of death and cremation ceremonies; this reflects the importance of the daily passage of the sun in relation to the human life cycle. On the neighbouring island of Lombok, the Sasak villages, although not as rigidly laid out, still reflect a similar world view. The dual nature of these villages is kept but individual houses' positions are influenced by local contours and placed atop hills, leaving the flat land below free for rice cultivation. The Dayak peoples of Kalimantan locate their villages along riverbanks within the dense jungle of the interior of the island. For them the duality of upstream and downstream are important determinants of village layout with upstream being associated with good spirits and downstream with evil spirits.

One of the most rigidly planned forms of rural settlement in Indonesia is to be found on the South Nias island off Sumatra. Here both the duality of opposites and social structure play important roles in village plans. The buildings, which house up to 2000 people, are arranged around a broad paved and stepped main village street. At the top end of this is the chief's house, positioned high above the rest. To the right of the chief's house is a row of longhouses known as the *abolata*, or sunrise, and to the left longhouses known as the *aechuta*, or sunset. Various stone ritual monuments such as columns and seats dedicated to the ancestors are placed in the main street outside the homes of important villagers; also in the main street are the ritual high jump stones over which villagers leap during important festivities.

The lower part of the South Nias village, where the poorer people live, is associated with the downstream elements of the cosmos and there are carvings of crocodiles and lizards, both of which are

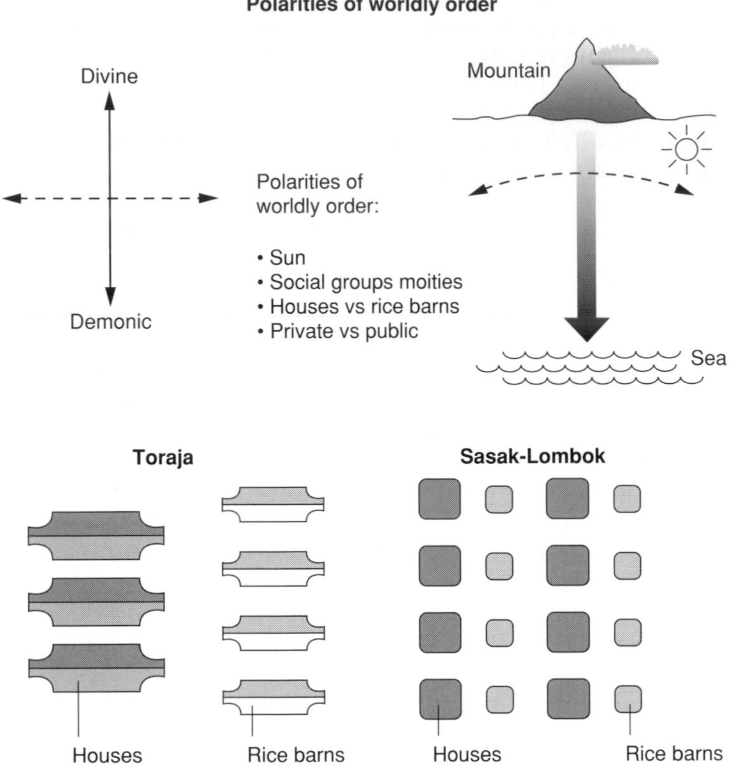

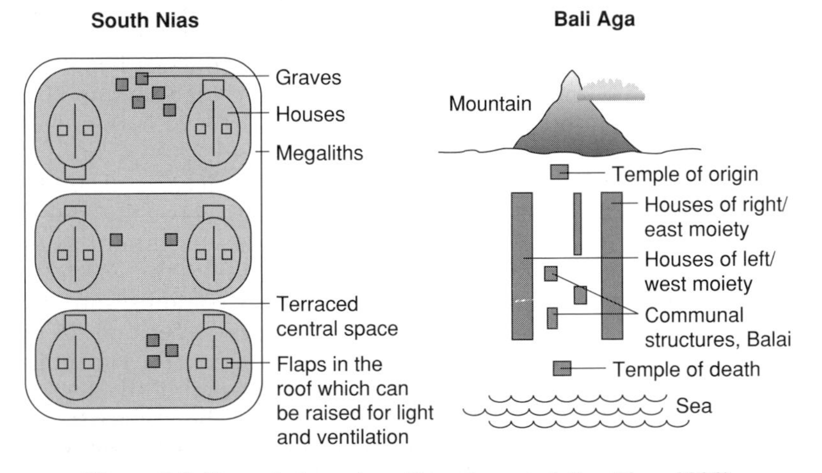

Figure 3.3 Some Indonesian village layouts (after Rigg, 1996)

regarded as demons. By contrast, in the upper parts of the village, where the higher status people live, the associated creatures are birds, symbolising the good spirits of the upstream world.

CASE STUDY: RURAL SETTLEMENT PATTERNS OF THE DANI OF WEST PAPUA

Outside of the mainstream cultures scattered through the Indonesian archipelago are the Dani of West Papua (until recently known as Irian Jaya). Despite contact with the outside world, many Dani villages in the remote Baliem valley of the central highlands of Papua have remained completely unaltered by Western influences.

Although the Dutch expeditions in New Guinea had made some contact with various tribal groups, the Dani were unknown to the outside world and their homeland was believed to be totally uninhabited, until US airmen spotted intricate field patterns in the middle of the dense jungle during World War II. The Dani, who arrived in the Baliem valley some 25 000 years ago, still essentially live in a Stone Age culture, despite having some metal implements, with an economy based on growing *ipere* or sweet potatoes and the rearing of pigs. There are some 30 clans that give a total population of about 75 000, which is growing rapidly as a result of their traditional ritual warfare having been almost totally abandoned and the effects of the medicines brought to them by missionary doctors.

Rural settlement of the Dani takes the form of compounds housing family units. Between two and five of these compounds are clustered together and are inhabited by families bonded by ties of clan; they are surrounded by the carefully tended, stone-walled and irrigated sweet potato fields. Each compound is laid out on a similar pattern. The compound is surrounded by a rectangular fence, partly defensive and partly to prevent the pigs from escaping into the potato fields. At one end of the compound is the large circular *honai* or men's hut, which is the main focus of communal meetings; along one side of the compound is a series of small round huts where the women live and opposite is the elongated cooking shed and covered pig stalls. The separation of men and women and some fairly strict taboos about when sexual intercourse is allowed were until recently reasons why the Dani population grew very slowly. The village layout closely reflects the social structure of the village; men are dominant and are allowed to practise polygamy. In the days when warfare between clans and cannibalism of enemies were widely practised, there were frequent raids on compounds to steal women and pigs; these were also the two main commodities that could be used as gifts to appease enemy clans.

Both Christian missionaries and the Indonesian government have tried to change the Dani lifestyle by trying to settle them in new villages made from wooden huts with tin roofs. Although schools and hospitals along the Baliem valley take this form, the rural Dani have largely resisted moving into similar settlements. Although Wamena, the valley's capital and service centre where the housing is of modern materials has seen some inward migration of Dani people, most of its inhabitants are *trasmigrasi* from Java and other parts of Indonesia. The traditional mud and grass built Dani homes are better equipped for the extremes of day and night temperatures experienced high up in the Baliem valley, but do have the disadvantage of not having proper chimneys; consequently large numbers of the Dani suffer from eye problems. The future of the Dani is closely bound up with the weakening political state of Indonesia. As such a distinctive culture, with closer ties to Papua New Guinea than Java, the West Papua independence movements have been fighting for their freedom for the last few decades and have been recently spurred on by the success of East Timor in breaking away from Jakarta's rule.

7 The Changing Morphology of Rural Settlements

Rural settlements are subject to changes through time, with various phases of rebuilding and restructuring. Changes are brought about by five major interrelated types of influences:

- demographic change through growth or contraction of village populations
- environmental factors either taking place gradually, e.g. climate change, or in a more cataclysmic way, e.g. earthquakes and landslides
- shifts in land use, whether this involves bringing more land under cultivation, or a change from one specialism to another
- political actions – decisions or upheavals taking place locally or nationally
- changes in land ownership.

In Britain the changing morphology of villages over the last 200 years can be studied from the evidence of accurate large-scale maps. This cartographical evidence comes from individual estate maps, enclosure maps, tythe maps and the various topographical editions produced by the Ordnance Survey.

Changing settlement morphology can be put into four categories:

(a) *Growth*. Villages and hamlets grow when they experience population increase. This growth may take a variety of forms, depending on the existing layout of a settlement. In linear villages or those with a radial pattern of roads, restrictions in the physical environment may force them to develop in a **linear** way. Where there is a lot of space within the existing village fabric, the new developments may take place through **infill**. Where there is land available on the periphery of a settlement, it may grow by **accretion**, i.e. the adding on of new housing around its edges.

Growth is either unplanned and 'organic' or planned; planning has become much more restrictive on settlement growth in Britain since the Town and Country Planning Act of 1947. In past centuries individual houses were added to the fabric of villages through **squatting** and **assarting**. Squatters were allowed to build cottages on common land and wasteland if they could put their buildings up overnight. Assarts were new plots of land cleared from forests on which cottages were frequently built.

(b) *Shrinkage.* This occurs where population decline sets in and may take a wide variety of forms from the loss of a few houses, through to whole streets or, in extreme cases, total abandonment of the whole settlement. The reasons for these changes are dealt with in Chapter 5.

(c) *Equilibrium.* Villages that have experienced little population change over hundreds of years may retain their historic plans intact. The main change that takes place within them would be the rebuilding, upgrading and extending of individual houses and farms. This situation is most common in hamlets and smaller villages that lack the population dynamics of larger rural settlements.

(d) *Replanning.* In a limited number of situations old settlements may be demolished and replaced by something totally new and better planned. During the seventeenth to nineteenth centuries in Britain many landowners were engaged in developing new rural settlements to replace existing ones. Examples of this are looked at in Chapter 5.

CASE STUDY: INFILL AND ACCRETION: THE DEVELOPMENT AND CHANGE IN THE MORPHOLOGY OF CHEDDAR, SOMERSET

Cheddar is a large sprawling village located on the south-facing slope of the Mendip Hills in Somerset (see Figures 3.4a and b on pages 54–5). The main slope elements on which it is sited are: the steeply rising slopes of the Mendip anticline, which are generally unsuitable for any large developments, the gentle colluvial footslope below the hills on which most of the village is built and the plains at the edge of the Somerset Levels upon which any developments are subject to the hazard of flooding.

The first modern topographical map of the village was an estate map drawn in 1788. This shows the village core as a 'Y' shape with the market cross as its focal point and the church at its southern end; elsewhere there are small farms and cottages scattered along roads and lanes. In between all this are large areas of farmland as shown in Figure 3.4(a).

Over the next 200 or so the morphological structure of the village changed beyond recognition into the densely packed agglomeration that it is today, as shown in Figure 3.4(b). The changes in Cheddar's morphology can be put into three main phases. The first phase was during the nineteenth century when the population grew relatively slowly. The main changes were the addition of houses along the existing roads and lanes, filling in the spaces between older cottages and farms, and this also caused the village to undergo some radial growth. Thus, in the nineteenth century the village changed little in its shape. Much of the open space within the settlement that had been arable field strips continued to have agricultural use as allotments, market gardens and orchards.

Throughout most of the twentieth century Cheddar saw a great deal of infill taking place within its internal open spaces, but the process was gradual. In the interwar period two areas of local authority housing were the most significant developments,

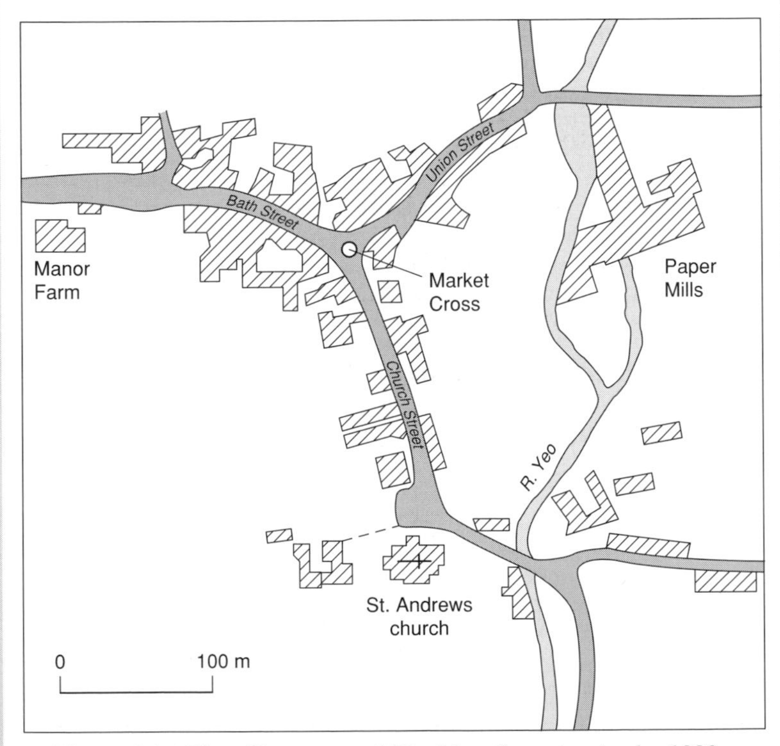

Figure 3.4a The village core of Cheddar, Somerset in the 1880s

the larger of the two being the Westacre estate with about 60 housing units. The process of infill accelerated between the 1960s and 1980s; the vast majority of this housing was built as private estates, the largest of which was the St Andrew's Road area of the village where over 80 units were built. By the mid-1990s there were virtually no open spaces left within the centre of the village. One area of prime building land to the north of the village, the Hamfield, remained untouched by development and still continues to be used for market gardening; it is one of the last relics of the open common field system

The third and most recent phase of housing development started in the late 1990s with building on the edges of the village,

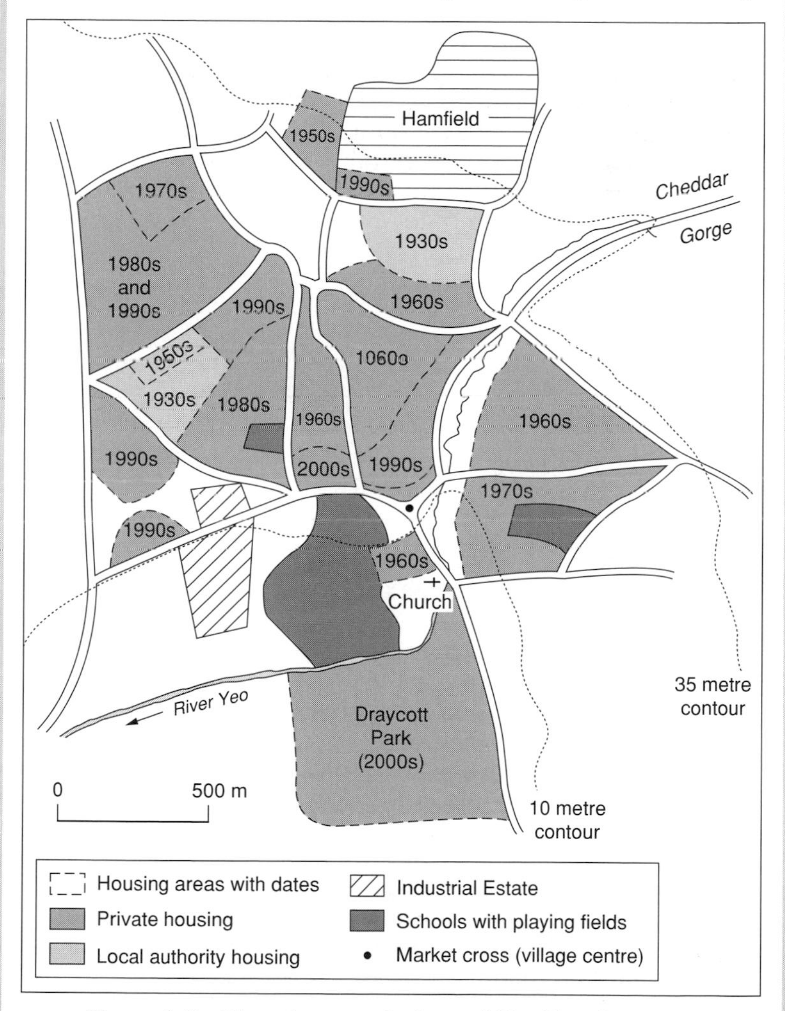

Figure 3.4b Changing morphology of Cheddar, Somerset

which is therefore now growing by accretion. The most ambitious of recent schemes that saw completion in 2003, is the Draycott Park development with 218 houses built on flat land close to the River Yeo. Although provision has been made for a floodwater spillway, it remains to be seen just how wise it was to build so many houses on this land.

The Draycott Park development in Cheddar alongside the River Yeo.

Questions

1. Explain why there are such variations in the layout of villages and other rural settlements.
2. How can the different forms of rural settlement layout be classified?
3. With reference to specific examples, examine the differences between the factors that influence rural settlement forms in MEDCs from those in LEDCs.
4. How may the morphology of a rural settlement change through time and what are the main causes of these changes?
5. In what ways does the natural environment influence the shape and form of rural settlements? Are physical factors generally more or less important than human ones?

Summary Diagram

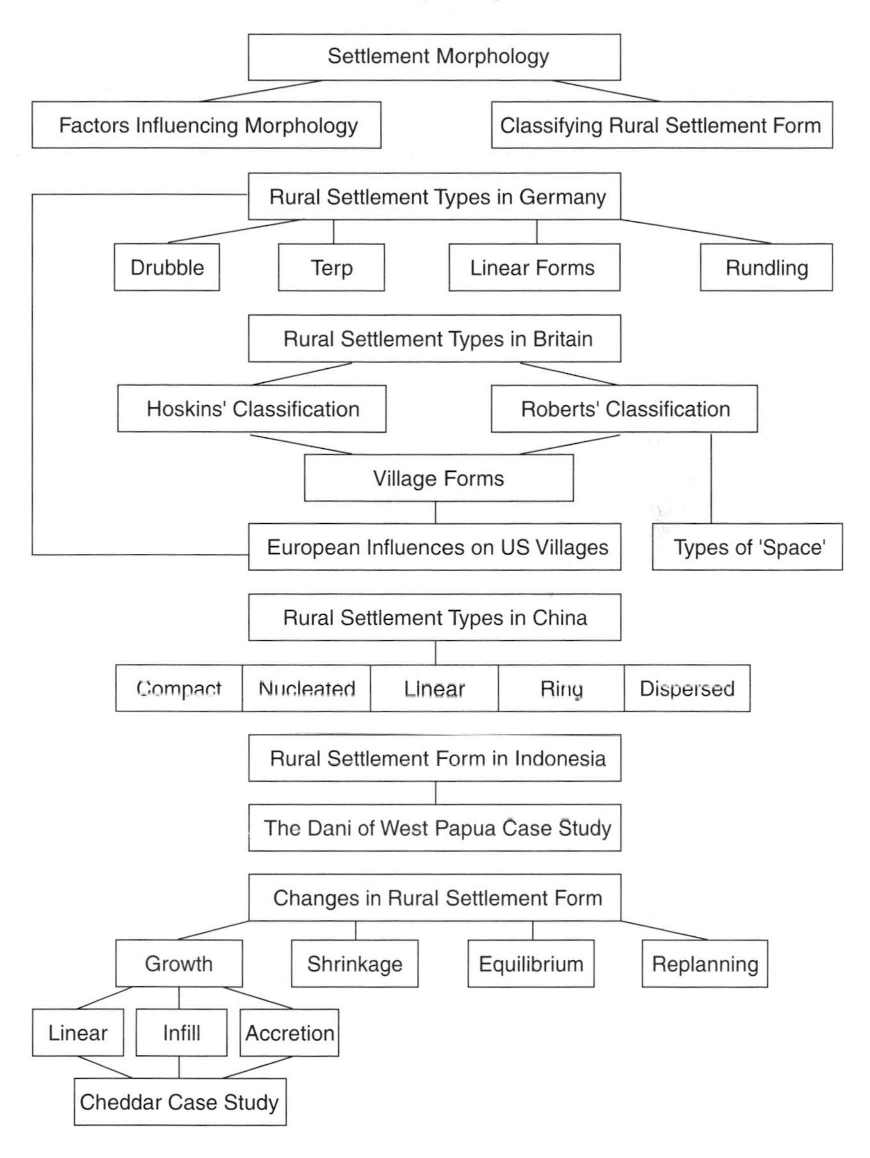

4 Rural Settlement Patterns and Hierarchies

KEY WORDS

Settlement hierarchy: the arrangement of settlements into categories according to size and rank.
Settlement density: the number of settlements per given unit area (e.g. per 10 km^2 or 100 km^2).
Isotropic surface: a landscape or other surface lacking topographical variations.
Settlement rationalisation: the choosing by a local authority or planners of 'key' settlements that are intended to become more important than their neighbours; the process of concentrating important functions in these 'key' settlements.
Manipulated hierarchy: a form of settlement rationalisation by which planners bring about alterations to pre-existing hierarchies.
Planned hierarchy: a settlement hierarchy that is designed and imposed upon a new landscape, e.g. a reclaimed marshland.

The town and village, dome and farm,
Each give each a double charm

John Dyer *Grongar Hill*

1 Nucleation, Dispersion and Other Patterns

The concepts of nucleation and dispersion were introduced in Chapter 1. As was stated in that chapter, nucleation is associated with village settlements, and dispersion with smaller scattered rural settlements including farmhouses and hamlets; the relationships between these two patterns of settlement will be considered below within the European context.

As well as this basic distinctions between nucleation and dispersal, geographers are also concerned with the **spacing**, **density** and **distribution patterns** of rural settlements within a given area. A whole series of interrrelated physical, social and economic factors influence the ways in which rural settlement patterns are spaced and distributed. As was seen in the section on settlement location, the fact that most rural settlements in the world have a long history means that these factors must be considered in a historic context rather than in the context of the contemporary world with its modern technology. The main influences on rural settlement distribution include:

- the fertility of the soil, including such factors as drainage: the more fertile the soil, the greater the population and settlement densities

- climatic conditions, especially rainfall distribution and temperature range, also influence the productivity of the land and therefore the settlement patterns
- major relief features such as hills, mountains, plains and valleys have a profound influence on settlement spacing
- accessibility, routeways and transport potential, and the economic activities they may stimulate
- the population density and the way in which it increases or decreases through time reflecting many of the physical factors mentioned above
- the need for defensive sites
- the existence of any antecedent settlement patterns.

Given the original agricultural nature of rural settlements, the fertility of the soil would have been the most important factor in determining the distribution and spacing of villages, hamlets or farmsteads. From the documentary evidence of the Domesday Book of 1086, the three most populous counties in England were Norfolk (with 95 000 people) Lincolnshire (with 90 000 people) and Suffolk (with 7000 people), all low-lying areas and endowed with relatively rich soils. Hoskins, when commenting on the spacing of Saxon villages in the highland and lowland zones of England, alludes to both soil fertility and antecedent settlement patterns:

> The Anglo-Saxons covered the whole of England with their villages, much more thickly in some parts than others. In Leicestershire and Lincolnshire, for example, the villages were often less than a couple of miles apart, and the Scandinavian settlement later added to the 'congestion'; but in Devon and Cornwall they were half a dozen miles or so apart, especially to the west of the Exe, probably because Celtic hamlets and farmsteads survived in not inconsiderable numbers and occupied much of the intervening country.

2 Nucleation and Dispersion in Europe

Some of the main reasons for nucleation and dispersion of rural settlements were dealt with in Chapter 1, as were the ways of measuring the degree of dispersion and the distribution of these patterns in Britain. Figure 4.1 shows how the patterns of nucleation and dispersal occur within the wider context of Europe. The patterns appear at first sight to be very complex, but on closer examination relationships can be seen between landscape types and the nature of their rural settlements. Mountainous regions and other peripheral areas are those in which the dispersed forms of rural settlement are dominant. These areas include the Alps, the Northern Apennines in Italy, the Massif Central and Brittany in France, the heathlands of northern Germany, Galicia in NW Spain and the central regions of the Balkans, especially

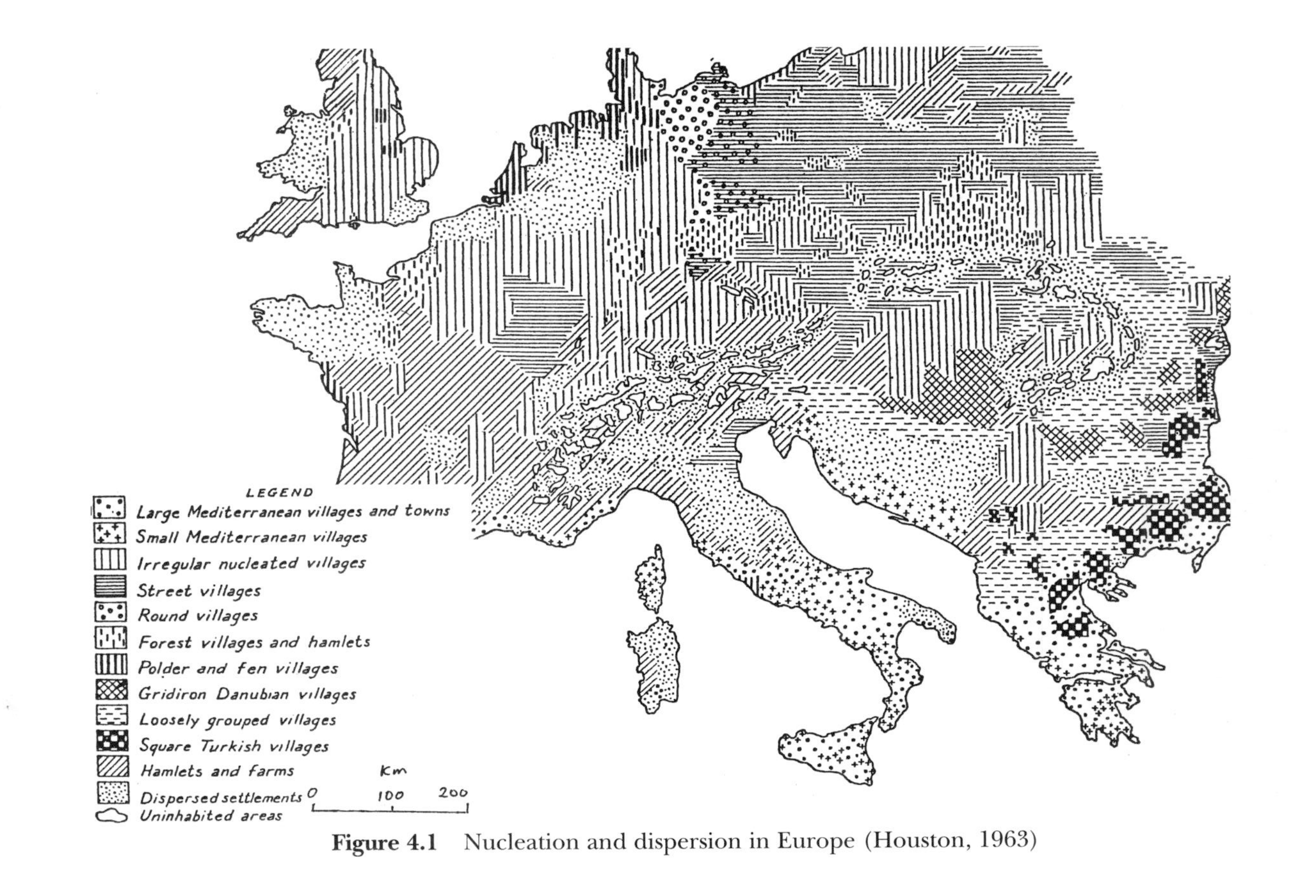

Figure 4.1 Nucleation and dispersion in Europe (Houston, 1963)

parts of Bosnia and Serbia. All these areas have low population densities because of their relatively difficult natural environments. Also within the dispersed settlement category are those areas with hamlets as the dominant settlement type. These include much of SW France, Liguria in Italy, Cantabria in N Spain, parts of S Germany and Lithuania, all areas that have relatively low population densities that once again reflect relatively difficult natural environments.

Nucleated villages dominate two types of region: those with rich soils and those with the historical need for defensive settlement sites. In the former category, alongside eastern and Midland England are the central parts of Germany, including the rich *loess* soil belt, and the Danube plains of Hungary. Those areas with large nucleated villages in defensive sites include southern Italy, including Sicily, Provence in southern France, most of Greece and the Dalmatian coastline of Croatia; all of these areas have experienced centuries of turbulence in the past. They also are located in parts of Europe that were previously highly susceptible to malaria and their hilltop locations would have offered some protection from the disease, which was found in lowland swamps.

CASE STUDY: THE IMPACT OF ENCLOSURE UPON RURAL SETTLEMENT PATTERNS IN BRITAIN

Changes in land use and in the productiveness of farmland can also alter patterns of settlement. The enclosure movements, particularly in the eighteenth and nineteenth centuries had a profound effect upon the patterns of nucleation and dispersion in Britain, especially in the lowland zone of England. Although there always were some isolated farmsteads and hamlets within the landscape of the lowland zone, from Saxon times the large nucleated village had been the dominant type of rural settlement within the Midland and eastern counties of England. Enclosure of common lands and wastelands took place in a piecemeal way over the centuries in some areas of England such as Kent and Devon, but in the most productive farmlands of the lowland zone, the open common arable fields with their strip cultivation continued intact until the pressures of commercial farming from the Georgian through to the early Victorian period led to the systematic enclosure of common land by Act of Parliament. The main landowner of each parish would instigate these profound changes. Common arable fields as well as meadow, moor, heath and any other type of wasteland or grazing land was enclosed and

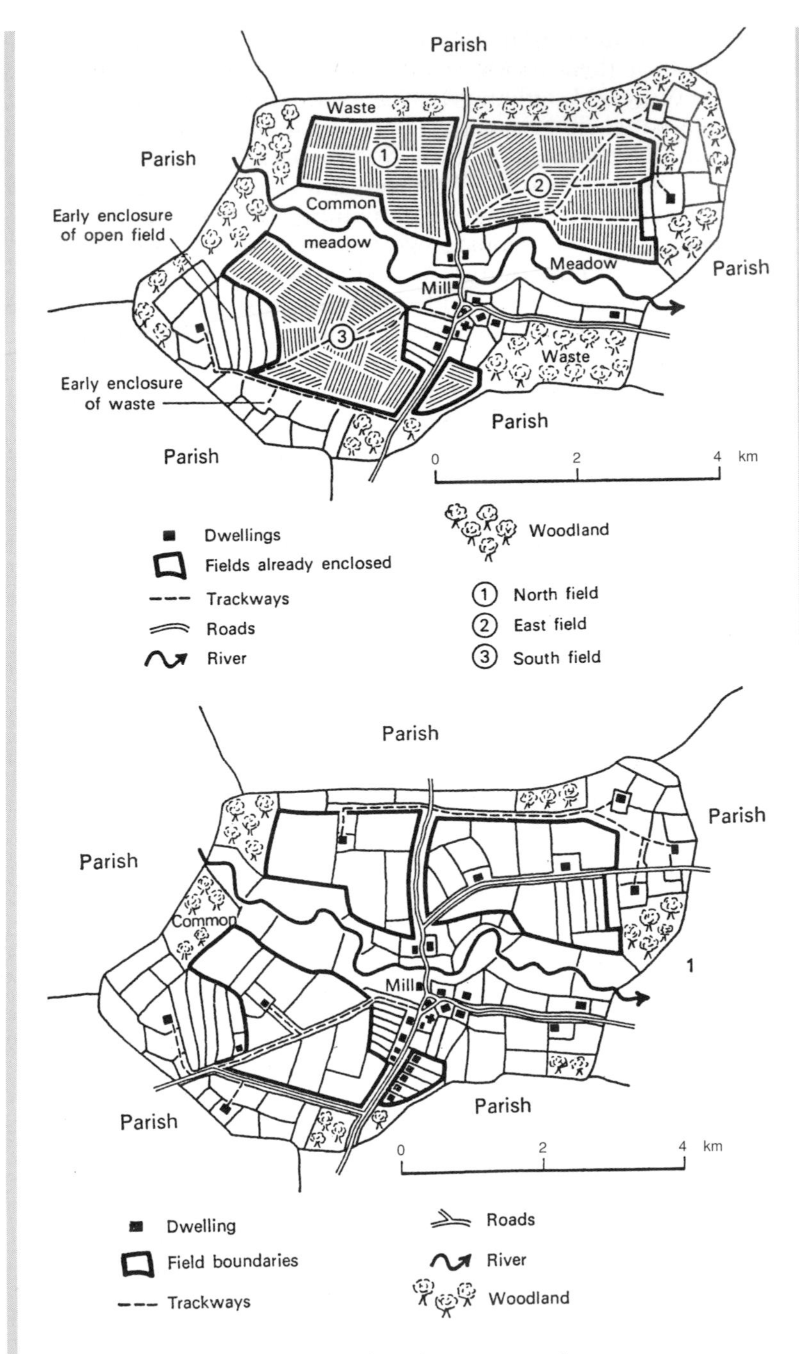

Figure 4.2 The impact of Enclosure on settlement patterns (after Everson and Fitzgerald, 1969)

the whole territory of the village divided up between a number of new farms. Figure 4.2 shows the impact this had upon the settlement pattern of a lowland village. Although the village cores of so many parishes throughout England may have numerous Mediaeval and Tudor buildings, the majority of farmhouses scattered around them belong to the Georgian Age. Likewise, so many of the hedges and walls that form the field boundaries are not ancient, but only around 150–250 years old. Place name evidence is a useful indicator of the period in which farmsteads were established, and although most of the new farms built at the time of parliamentary enclosure took on local names, occasionally their names reflect places, people, events and aspirations of the Georgian Age. 'Pennsylvania', 'Wellington', Waterloo' and 'Providence' are all examples of farm names dating from this period.

In the Mendips in north Somerset 23 parishes were affected by Acts of Enclosure between 1770 and 1854 – although only three of these took place after 1800. A few parishes in the east of the region had already enclosed their common land before this main period of enclosure without parliamentary approval. These villages were small and located in gently rolling countryside. The central and western parts of the Mendips had much more dramatic scenery to tame: high plateaux, steep slopes, rocky valleys – and the parishes on the southern slopes extended down into the marshlands of the Somerset Levels. To improve, reclaim and enclose these parish territories involved great expense and effort, and therefore an Act of Parliament was required to authorise the work. Prior to the Enclosure Acts each county was reported upon by the Board of Agriculture as to what changes could be made in order to enclose the land and thereby make farming more commercial. In the case of Somerset, the survey was carried out and written up by John Billingsley, a major landowner in the eastern part of the Mendips. By the time of the survey, enclosure was already underway, but what is noticeable is that parishes in the east were generally enclosed earlier than those in the west and there was a gradual westward diffusion of the process across the Mendips. The way in which enclosure progressed across the Mendips is no doubt due to both Billingsley's influence and to where he owned land. The Acts of Enclosure within the parishes of the Mendips led to the establishment of three types of new farms and landscapes:

- within the villages themselves new Georgian farms were established upon land that had been open common cereal fields; their new field boundaries were generally hedges and trees that by now have reached great maturity
- on the top of the Mendips the new Georgian farms were more extensive as they were established on lower quality soils of the

windswept plateau; their field boundaries were generally dry limestone walling

- on the reclaimed marshland of the Somerset Levels the new Georgian farms were devoted to medium intensity dairying; their field boundaries were the straight ditches known locally as *rhynes*, which were to protect the land from winter flooding.

Enclosure not only added to the existing rural settlement patterns, but in some places led to the establishment of a totally new pattern of settlement. The colonisation of one of the highest parts of Exmoor in west Somerset was one of the great farming achievements of the nineteenth century. In 1815 an Act of Enclosure was passed to enclose the lands of the old Royal Forest of Exmoor. Most of this upland moorland, which was devoid of trees, roads, hedges or farmsteads, was acquired by the Knight family who transformed it into an area of productive farmland. The operation was centred upon the small hamlet of Simonsbath, which had been established in 1654 and was expanded by the Knights. Between 1825 and 1861, 16 new livestock farms with areas between 100 and 1000 ha were established and this part of Exmoor became a landscape of farms, fields, hedges, trees, sheep and country roads: a similar farmscape to much lower lying areas (see Figure 4.3).

3 Patterns of Spacing and Density of Rural Settlement

Figure 4.4 shows the main types of patterns that rural settlements may form. There is first the two main contrasting forms of **regular** and **random** spacing. Patterns are more likely to be regular in areas where there are few contrasts in topography, best typified by plains and where there has been some degree of planning of the settlement pattern. Regular spacing rarely occurs in reality as natural resources and other factors intrinsic to the siting of villages hamlets and farmsteads are seldom to be found evenly distributed throughout an area of countryside. Random spacing is therefore the norm for rural settlements and regular patterns are much rarer. The two most common variations upon these basic patterns are due to **clustering** and the influence of **linear** factors. Clustering may be created by a variety of factors, but is most likely to occur where there are more favourable conditions in some places, which are absent in others; variations in soil quality (e.g. fertility, depth, drainage, acidity) are perhaps the most significant factors in determining the clustering of settlements. The linear element to settlement patterns is created by strong topographic or human trend lines within the countryside; restricted valleys,

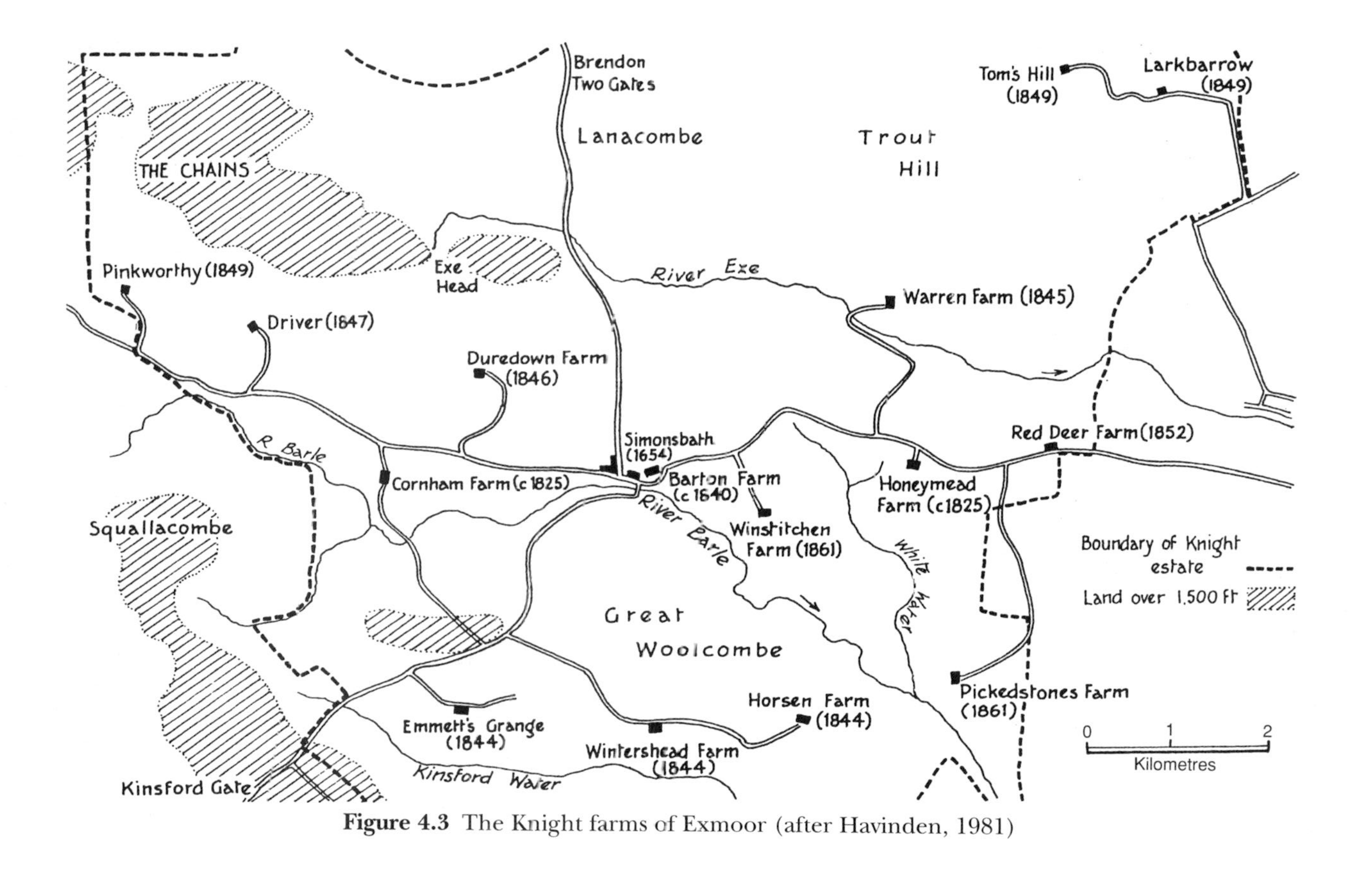

Figure 4.3 The Knight farms of Exmoor (after Havinden, 1981)

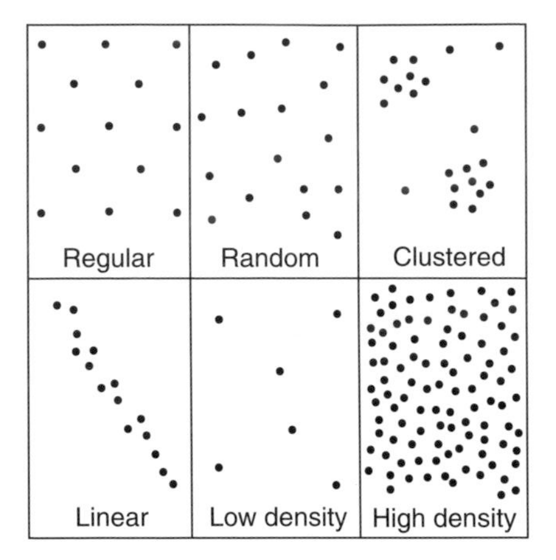

Figure 4.4 Settlement distribution patterns

major rivers, causeways through marshes and transport arteries are all responsible for such linear patterns. The difference between the random and regular elements within clustered and linear patterns are once again likely to be connected to the degree of human planning of the landscape and its settlements.

CASE STUDY: CONTRASTING RURAL SETTLEMENT PATTERNS IN ITALY

The patterns of rural settlement outlined above can be found in almost any part of the world. Figure 4.5 shows the distribution of nucleated rural settlements in five different parts of Italy. All five maps are at the same scale. The distribution of villages around the three provincial capitals of Cremona, Lecce and Matera reflect very different levels of agricultural potential. Cremona is located on the Pianura Padana (North Italian Plain) close to the confluence of the rivers Po and Adda. The Plain is the biggest area of flat land with fertile soils in Italy and therefore supports a relatively dense rural population. The absence of hills or mountains has enabled a relatively uniform pattern of population distribution and therefore settlement distribution to evolve. The only place shown on the Cremona map to have a slightly lower density of villages is in the south-west along the actual valley of the Po, which is already in the flood plain stage and therefore presents a natural threat to human habitation.

Lecce is located on the Salentine Peninsula in the eastern part of the Puglia region in the 'heel' of the Italian peninsula.

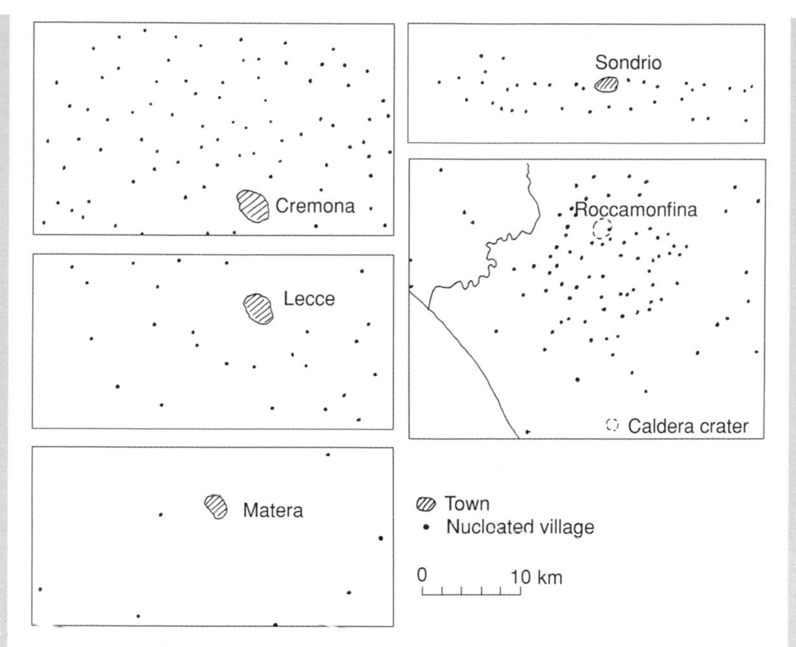

Figure 4.5 Italian nucleated settlement patterns

The villages in this part of Italy are much larger than those of the Pianura Padana, therefore the population density is actually higher in Lecce province (289 people per km²) than in Cremona province (187 people per km²), despite there being fewer and more widely spaced settlements. The Salentino is one of the few flat areas in southern Italy, and has rich terra rossa soils developed upon the local limestone. Water supply is the one major problem of the area, but this has largely been overcome by the construction of the Puglian Aqueduct network of pipelines that has been gradually extended over the last 80 years. The region therefore has considerable agricultural wealth from high value crops such as grape vines, tobacco, citrus fruits, olives and almonds, which enables it to support a dense rural population. The distribution pattern of villages is slightly less even than in the Cremona area as there are more local variations in topography, such as dry valleys within the limestone.

Matera province in Basilicata has a population density of just 60 people per km². As in the Lecce province, the villages are very large nucleated ones, but there are far fewer of them. Matera province is mainly barren hill country made of infertile clays, shales, limestones and sandstones. Overexploitation of the poor soils over the centuries has led to vast areas of gullied lunar landscapes and the region is much less productive than the Cremona or Lecce regions. This is reflected in the very scattered pattern of rural settlement. The villages in this region are located on hill-tops overlooking the major valleys and this accounts for their distribution pattern.

Sondrio is the provincial capital of the Valtellina region located in the Lombardy Alps. Valtellina is, in fact, a deep lateral glacial trough trending from east to west through the Alps and contains the upper reaches of the River Adda. Topography imposes great restraints upon both agriculture and settlement and therefore the pattern of village distribution is clearly linear. Villages are located on both sides of the Valtellina, above the river on small hills, terraces or alluvial fans. There is a marked difference in the sizes of settlements due to aspect; the villages on the south-facing slopes are considerably larger than those on the north-facing slopes. Some of the north-facing slope villages lack the direct rays of the sun for several months in the year. Sondrio province has a population density of only 55 people per km^2, but of course this is much higher in the valley itself. For a peripheral rural area the Valtellina is relatively wealthy because of its high-grade agricultural products (e.g. some of Italy's very best wines) and because of the winter sports industry.

The Roccamonfina area to the north of Naples in the Campania region of Italy provides a good example of a clustered rural settlement pattern. Much of the zone between Rome and Naples is dominated by either coastal plains or high limestone mountain ranges, both of which are sparsely populated. Roccamonfina is the caldera of a former strato-volcano that has remained inactive for at least 2 million years. The volcanic ash has broken down into soils that are much richer than those of the surrounding areas and have therefore long supported a denser agricultural population, reflected in the degree of clustering of villages on the mountain's flanks. The villages on the slopes of Roccamonfina are also historically well located for defence from coastal invasions as well as being elevated above the malarial areas of the coastal plain.

4 Christaller and the Central Place Theory

Many of the present-day concepts of settlement hierarchy have developed out of the theories put forward by Walther Christaller in Germany in the 1930s. His extensive studies of the relationships between settlement sizes, their range of services and the distances between them led to the publication of his seminal work in 1933: '*Die Zentralen Orte in Süddeutschland*' ('Central places in southern Germany'). Christaller identified a hierarchy of seven different

Table 4.1 Settlement hierarchy in north-east England

Grade 1 Centre:	Regional capital
Grade 2 Centres:	Five settlements: 4500 to 12 500 adults
Grade 3 Centres:	23 large market settlements: 900 to 4500 adults
Grade 4 Centres:	33 medium market settlements: 411 to 900 adults
Grade 5 Centres:	56 agricultural service settlements: 165 to 410 adults
Grade 6 Centres:	58 small villages: 90 to 164 adults
Grade 7 Centres:	130 hamlets: 36 to 89 adults
Grade 8 Centres:	Small hamlets and farmsteads: less than 36 adults

After Edwards.

categories of settlements, from the *Landhaupstsadt* or state capital (Munich) at the top, down to the village at the bottom. At each stage in his hierarchy there are close relationships between the settlement size, the number of settlements in that category and the distance between them. At the top of the hierarchy there was only one settlement, which had a population of 500 000, and it was 186 km from its nearest equivalent in the neighbouring state. At the bottom of the hierarchy Christaller identified 486 *Marktort* settlements (small market centres), which had an average population of 1000 and were spaced on average 7 km apart. The theoretical models which Christaller produced from his research showed how on an **isotropic surface**, i.e. an area of land with no topographic variations, settlements in each category would be regularly spaced and have hexagonal spheres of influence or market areas.

Christaller's theory has been greatly discussed and adapted by geographers. Although it does not put much emphasis on rural settlements, Edwards, working in north-east England in the 1970s produced a version of the Central Place Theory that related much more closely to the rural environment. As can be seen from Table 4.1, the Grade 3 Centres through to the Grade 8 Centres are all relevant to the study of the relationships between rural settlements and the marketing of agricultural produce.

The main criticisms of the Central Place Theory revolve around the differences between theory and reality, and how it oversimplifies what exists in the real world and assumes conditions that do not exist in reality. In its defence, however, it can be argued that as with other theoretical models in geography, its purpose is to explain in a simple way the basic principles that underlie a much more complex reality. One way in which the real landscape deviates from that of the Central Place Theory is in the number of different hierarchical levels, which are found within any area. Even within one region of a country there may be great variations in settlement hierarchy patterns, as can be seen in the following case study.

CASE STUDY: CONTRASTING RURAL SETTLEMENT HIERARCHIES IN PUGLIA, SOUTHERN ITALY

Figure 4.6 shows three contrasting patterns of rural settlement within the Italian region of Puglia. In each case there is a different combination of rural settlements, resulting from both differing environmental and historical influences. Puglia itself is located in the 'heel' of the Italian Peninsula and, although a region composed almost totally of limestone, it has considerable variations in relief and soil quality and therefore in land use.

The first of the three areas shown on the map has a hierarchy dominated by large nucleated villages, which have just a very few isolated farms in between. This is an area of poor soil quality associated with the landscape of the Murge – a series of naturally occurring broad limestone terraces that rise up from the Adriatic coast to almost 700 metres at their highest point. Population densities are low, extensive cultivation of Mediterranean tree crops such as olives and almonds is carried out on the lower slopes, and on the upper parts of the Murge land use is either extensive cultivation of *tritticum durum* (pasta wheat), rough pasture or wasteland. The poor land quality of this area is reflected in the settlement patterns. Within the Murge large nucleated villages have been traditionally the dominant settlement unit, as they are throughout most of southern Italy, but they are both smaller and more widely separated on the higher land than in the area closer to the sea.

The second pattern is dominated by dispersed settlements, with just a few large nucleations in between, which is quite rare in southern Italy. This is the Valle d'Itria area of Puglia where both the soil conditions and the cultural history help to explain the settlement pattern. The soils are deep, rich limestone developed **terra rossa**, which support very intensive cultivation of vines, figs and vegetables. There is a dense rural population in the area living in hamlets where the houses are in the traditional *trullo* form. These are whitewashed farm complexes with a series of characteristic conical roofs and are unique to the region. The *trulli* are likely to date back long before Roman times and therefore represent the continuity of a sophisticated and practical form of Iron Age settlement that has not been superseded by nucleated villages.

The third pattern, found on the Salentine plain around the port of Otranto, is a much more balanced one. It is a relatively low-lying area (under 90 metres above sea level) and has moderately good soils producing a wide range of Mediterranean crops such as vines, tree fruits, vegetables and tobacco; this in turn supports a medium density population. This landscape is the closest of the three to an isotropic surface and it is not surprising therefore that it has the nearest settlement pattern to that suggested by Christaller.

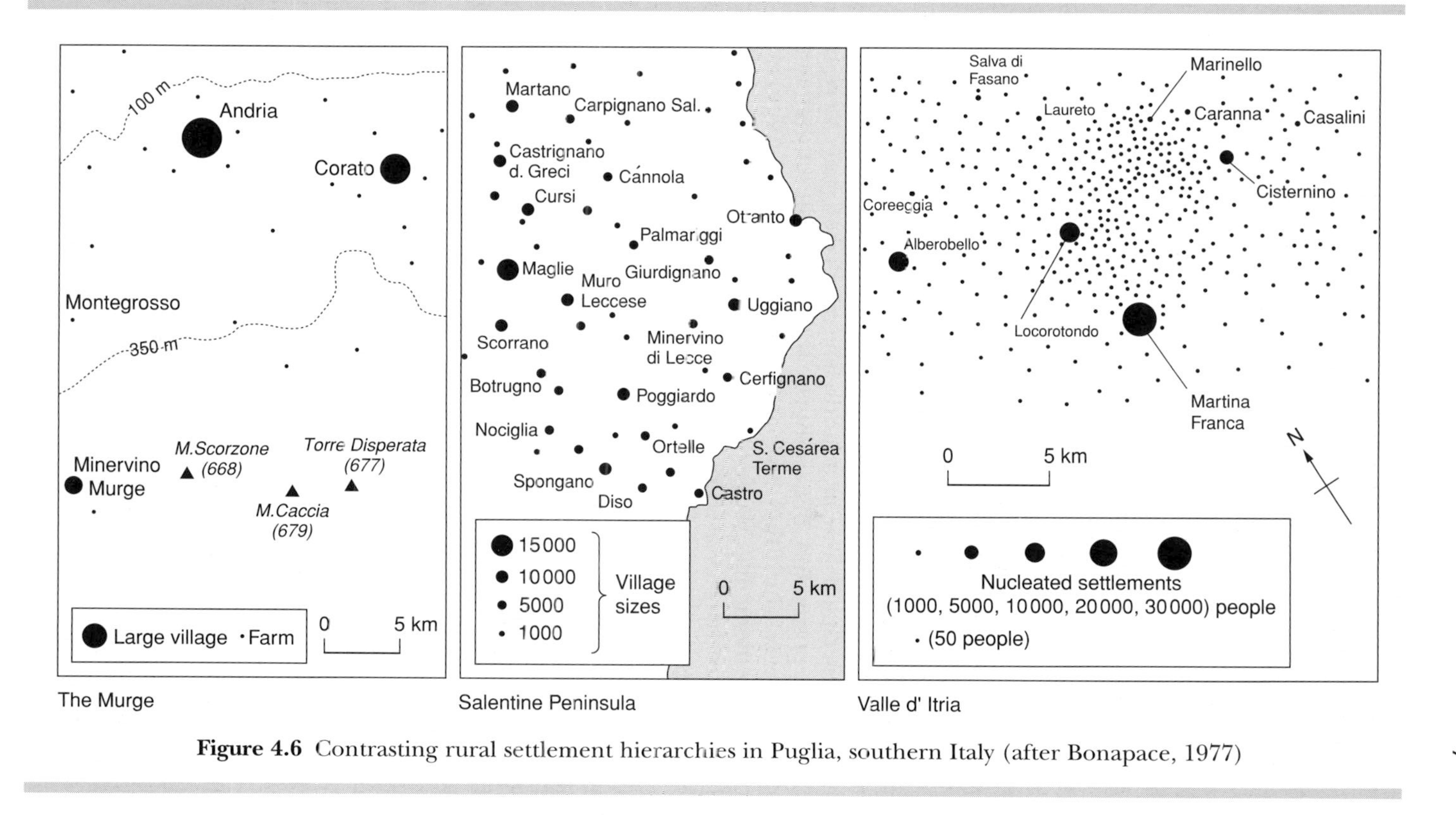

Figure 4.6 Contrasting rural settlement hierarchies in Puglia, southern Italy (after Bonapace, 1977)

5 The Evolution of Rural Settlement Hierarchies

Settlement hierarchies develop in a number of different ways: they can be said to 'evolve naturally' over a long period of time, they may be rigidly planned when an opportunity arises to create a new settlement pattern from scratch, or they may fall between the two when human interference can influence the way in which they are allowed to evolve. The parts of southern Germany studied by Christaller are a good example of a region where the hierarchy has evolved 'naturally' without much human planning or interference. In Britain, East Anglia has been singled out as a region with a fairly regular settlement hierarchy; as with South Germany, it has few great topographical variations. Examples of hierarchies that have been altered by human interference and imposed by rigid planning are given as case studies at the end of this chapter.

Exactly how the earliest rural settlement hierarchies evolved is lost in the past. In some societies political power may have been the deciding factor, in others it may have been just a matter of economics. It is easy to understand how in a landscape of villages some were more favourably positioned than others, both in terms of the natural resources over which they had control and the local route networks between settlements. Many of the first market towns throughout Europe must have evolved in this way from the village locations that had the best economic potential within their neighbourhood.

In Saxon and Mediaeval England the granting of royal charters that allowed settlements to hold fairs and markets was a formalised way in which villages could acquire town status. Thus, the king's patronage became a force behind the development of the settlement hierarchy. It was not necessarily the most economically viable places that became market centres in this way because borough status could be awarded as a reward to the local landowner for giving the king loyal support.

Hamlets should not necessarily be regarded as being lower down the hierarchical ladder than villages. In Cornwall there evolved a form of settlement known as a 'federal parish'. This has all the functions of a nucleated village, but they are separated out between a number of different but linked settlements. Typically, a federal parish would have within its boundaries three separate hamlets: one containing the church, another the manor house and the third one having its market functions.

CASE STUDY: THE SETTLEMENT HIERARCHY OF SEDGEMOOR DISTRICT, SOMERSET

Settlement hierarchies evolve through economic forces but can be changed by the deliberate action of local or national governments working on the advice of town and country planners. The process, which changes the relative importance of selected settlements, was referred to in the 1960s and 1970s in Britain as **settlement rationalisation**. Today this expression is rarely used as it has connotations of some settlements being favoured and others losing out, and is therefore too emotive. Since the 1960s many local authorities in the UK have drawn up plans that have selected key rural settlements and made them into **growth centres**. These settlements were given priority over other villages for the expansion of housing developments, new services and small-scale industries.

Local authorities have approached these changes in four main ways:

- **The market town approach**. This involves concentrating developments in existing towns rather than introducing much change to rural settlements. (This approach has been used in Cornwall and Herefordshire.)
- **The place-specific approach**. This involves the identification of specific places within a hierarchy and to make them into key villages in which the main developments in housing, employment and services would be concentrated (e.g. Somerset and Cambridgeshire).
- **The restraining policy**. This involves restricting most the growth within an area but allowing it to take place in a few selected locations (e.g. Essex and Kent).
- **The area approach policy**. This strategy does not select specific places but aims at making development well dispersed throughout the whole area, district or county (e.g. Cumbria and Gloucestershire).

In the process of rationalisation many of the smaller settlements that were not selected as growth centres lost certain functions such as schools and doctor's surgeries as they were merged and relocated in the key settlements. In this way a hierarchy that may have taken centuries to evolve, becomes radically altered. Therefore this type of settlement pattern can be referred to as a **manipulated hierarchy**.

Figure 4.7 shows the hierarchical structure of the Sedgemoor District in Somerset from its most recent plan of 1999. Bridgwater, the administrative centre of the District, together with Burnham on Sea and Highbridge are all urban centres with different but wide-ranging services and represent the top level of the hierarchy. The rest of the nucleated settlements within the Sedgemoor District are villages. Four of these (Cheddar, Wedmore, Nether Stowey and North Petherton) have been given the status of **rural centres**. They were chosen, to quote the Sedgemoor District plan:

> ... because of the range of facilities they offer and their location [they] serve as centres for their surrounding rural areas. Cheddar has a good range of facilities and is a local employment centre, North Petherton has a good range of services. Nether Stowey and Wedmore are smaller with less (sic) facilities, but because they are in relatively inaccessible parts of the district they are important local centres.

Sedgemoor Council is therefore selecting places that had to some degree evolved above the sizes of the other 48 villages and giving them a more elevated status. Not only do the service centres have a bigger range of services than the other villages, but they are also growth points where the majority of non-urban housing is going to be located within Sedgemoor District. Housing and other developments are very highly restricted in all the other villages. This means, therefore, that Sedgemoor District has managed to create a manipulated hierarchy, using a combination of the place-specific approach alongside a restraining policy.

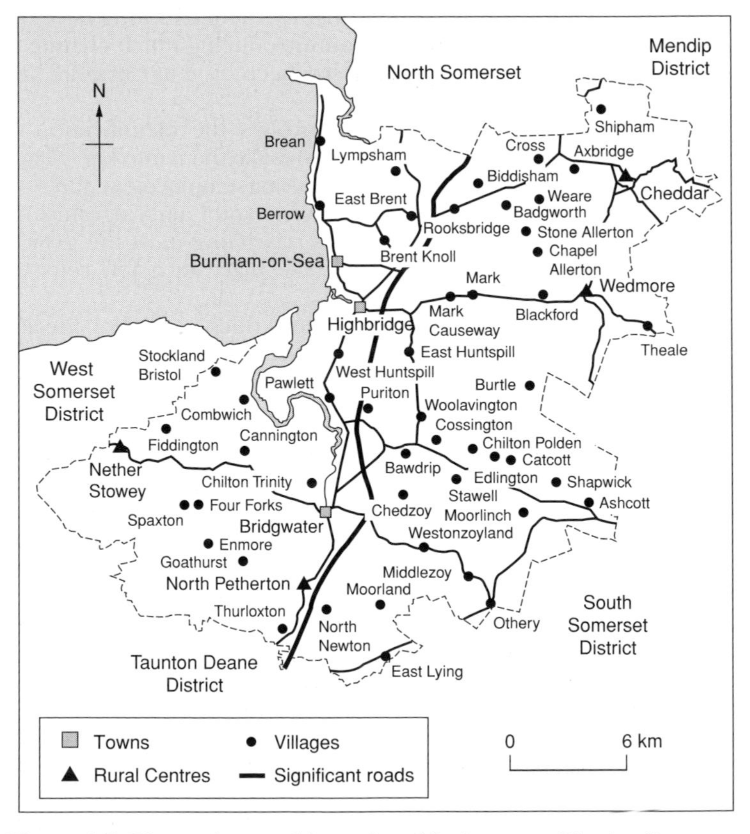

Figure 4.7 The settlement hierarchy of Sedgemoor District, Somerset

CASE STUDY: THE PLANNED SETTLEMENT HIERARCHY OF THE AGRO PONTINO, LAZIO, ITALY

The best examples of fully planned settlement hierarchies are to be found in areas that were newly colonised at some recent stage in history and were given their hierarchy by their colonisers. Throughout the parts of the tropics that were colonised by the European powers, there are areas of land where plantations or other types of farms were set up and run by European companies or individual coloniser. Hierarchies that have been established in this way include regularly placed farms with villages between them set up to act as service centres (e.g. the French colonial estate farms in northern Tunisia) and plantation processing villages with smaller settlements built around them to house the agricultural labourers (e.g. the tea plantations of central Sri Lanka).

Within Europe the most regular rural settlement hierarchies are to be found in areas that were reclaimed from marshlands within the last 200 years. Parts of the Fens in eastern England and the Somerset Levels have regularly spaced farmsteads upon them as a result of the reclamation and planning of the late eighteenth and early nineteenth centuries. The Polders of Holland have well-structured settlement patterns resulting from the reclamation schemes completed in the late twentieth century.

The Agro Pontino, the former Pontine marshes located to the south of Rome, have a well-defined settlement hierarchy that was established in the 1920s and 1930s by the Fascist government under Mussolini. The region that is located along the Tyrrhenian coast between Rome and Naples was for centuries an inhospitable and malarial marshland, used only for grazing land for transhumant sheep in the drier summer months. Attempts at reclamation of the Pontine Marshes by the ancient Romans and various Renaissance Popes were largely unsuccessful because of the malaria.

The marshes were successfully drained using twentieth century technology between 1926 and 1935, and constituted part of Mussolini's 'Battle for the Grain' (a series of measures to improve Italy's food production). Most of the work was carried out by the *Opera Nazionale Combattenti*, a Fascist war veterans' association. Altogether the area saw the reclamation of over 150 000 hectares of farmland, 800 km of new roads, 500 km of new canals and 18 pumping stations to regulate water levels. At the same time the reclamation work enabled in 1934 the setting up of one of Italy's earlier national parks, the *Parco Nazionale Circeo,* which covers some 7400 hectares and includes habitats as diverse as the limestone headland of M. Circeo itself, large tracts of Mediterranean oak forest and a sequence of coastal dunes and lagoons.

Upon the chequerboard of reclaimed land was placed a new settlement hierarchy. At the top level was the provincial capital, Latina that now has over 100 000 inhabitants. Below this in the hierarchy,

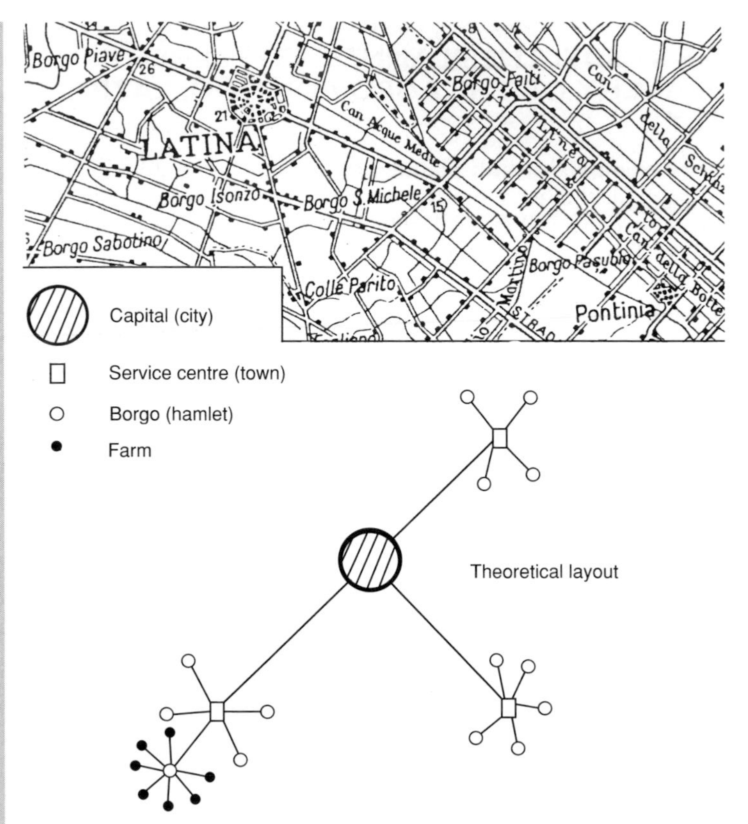

Figure 4.8 The settlement hierarchy of the Agro Pontino, Lazio, Italy

four new small towns were built to act as service centres; two of these, Pontinia and Sabaudia, were located on the Agro Pontino itself, the other two, Aprilia and Pomézia, were located on the road between the region and Rome, and have consequently become industrial centres. There are no real nucleated villages in the region as the next level down in the hierarchy is made up of 24 *borghi* or hamlets. These were nearly all new at the time of reclamation although some are centred on older settlements. These agricultural hamlets are composed of a few farms, a number of convenience stores, a church and the offices of agricultural cooperatives. At the bottom of the hierarchy are the 3500 individual farmsteads that were rigidly distributed throughout the new rich farmland as smallholdings.

Although this pattern is not as perfectly clear as when it was first established because of the consolidation of holdings, merging and subdivision of plots, as well as the building of holiday homes and urban sprawl from Latina and the other towns, it remains a classic example of the effects of settlement hierarchy planning (see Figure 4.8).

Questions

1. What are the different types of distribution patterns that are found in rural areas and what factors are responsible for them?
2. With the aid of specific examples explain how and why settlement hierarchies develop.
3. Examine the concept of the Central Place Theory and examine its relevance to real settlement patterns.
4. Why are some rural settlement patterns more regular than others?
5. With reference to specific examples examine what is meant by the terms 'manipulated' and 'planned' settlement hierarchies.

Summary Diagram

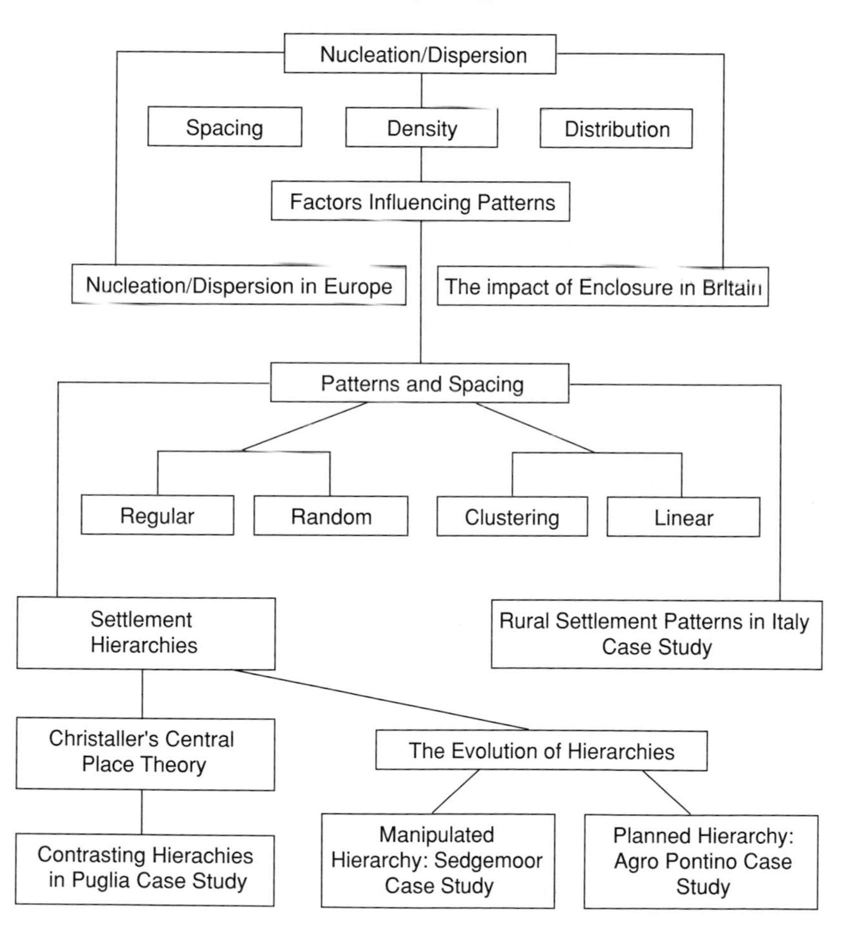

5 Rural Depopulation and Urbanisation

KEY WORDS

Rural depopulation: the migration of people away from villages and rural areas to towns.
Mobility Transition Model: a theoretical temporal model devised by Zelinsky that puts major migratory flows into a series of historical stages.
DMV (Deserted Mediaeval Village): one of several thousand English villages abandoned at the time of the Black Death or later.
Urbanisation: the rapid increase of the proportion of a country's or region's population living in urban areas as a result of rural–urban migration.
Step migration: a movement of people that take place in a series of spatial steps or stages.
Green belt: a designated area of open countryside, farmland and recreational land surrounding an urban area in which building is either forbidden or highly restricted.

> Yes this field used to be a village. My grandfather could call to mind when there were houses here. But the squire pulled 'em down, because poor folk were an eyesore to him
>
> Thomas Hardy *The Trumpet Major*

1 Reasons For Rural Depopulation

The clearing of villages and the expulsion of its peasant population by a ruthless landowner was a common occurrence over several centuries in Britain; this exemplifies one of many different ways by which rural settlements may become depopulated. Some settlements are more fortunate than others. As seen in the last chapter, villages that are well situated are more likely to evolve into market centres and then to work themselves up the 'hierarchical ladder' than those which are not so well positioned. In much the same manner, villages that are less fortunate in their locations may, for one reason or another, lose population and then eventually become abandoned altogether. Throughout history and from continent to continent, there has been a very wide range of reasons why rural settlements have become abandoned; many of these reasons are still valid explanations in the contemporary world. In Europe alone there are hundreds of thousands of rural settlements which were abandoned at various times in history. Among the many reasons for their desertion are:

- natural disasters such as earthquakes and landslides
- diseases and pestilence such as malaria and the plague
- destruction by invaders or during civil wars
- climate change and loss of agricultural potential
- overworking of the land, soil erosion and land degradation
- economic changes leading to new types of land use
- the actions of individual landowners.

Throughout the world, far more villages and other rural settlements have been affected by the less severe forms of rural depopulation, which involve a flow of migrants away from the individual settlement, rather than total abandonment. Typically, this form of migration involves the movement of people to a larger place in search of work or a better life. This type of gradual depopulation can have very dramatic effects upon villages, leaving them without an effective and capable workforce, setting them into a downward spiral of economic decline. Throughout LEDCs, the current process of urbanisation is resulting in heavy rural depopulation which takes many of the most able and educated people away from their villages. Rural depopulation and rural–urban migration as a process is best seen and understood in the context of the **Mobility Transition Model** which was devised in 1971 by the US demographer, Wilbur Zelinsky. He classified different types of migration according to a country's phases of economic development (Figure 5.1). In the first and pre-industrial phase, rural–rural migrations dominate, this being a reflection of

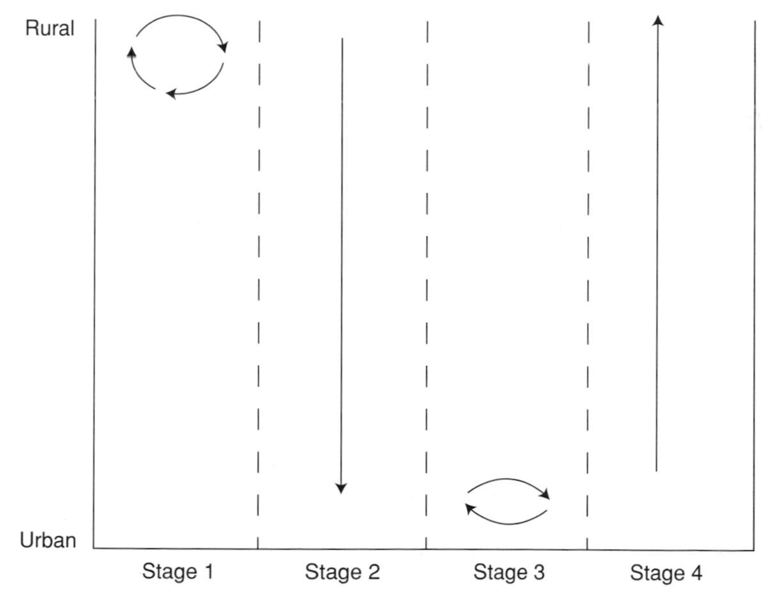

Figure 5.1 Zelinsky's Mobility Transition Model

what happens in rural societies that have either mobile farming systems (such as nomadic herding) or a large itinerant rural labour force. In the second phase, rural–urban migration is dominant; urbanisation and rural depopulation are taking place. This is the phase of industrialisation that took place in Europe in the nineteenth century and is happening in most LEDCs today. The third phase represents the situation in industrialised societies where most movements of people are between or within urban areas, particularly through commuting, therefore urban–urban migration waves are dominant. Zelinsky's Model's final phase relates to counterurbanisation and the movement of people from urban to rural areas, as is being experienced in most MEDCs today. The model therefore shows that rural depopulation is merely one of a sequence of stages through which all countries should theoretically go as they evolve economically. It can thus be concluded that what is happening in North America and Europe today should occur in other parts of the world in the near future, and indeed there is already some evidence that counterurbanisation is taking place today in and around some of the richer cities in LEDCs.

2 The Historical Sequence of Rural Settlement Abandonment in Britain

A study of the historical sequence of rural settlement abandonment and desertion in Britain gives great insight into the variety of reasons why specific rural areas became depopulated during certain phases in time. For the earliest periods of settlement, the evidence is both sketchy and incomplete. Archaeological evidence depends upon which sites have been discovered and excavated, and in many places is not available because of continuity of settlement on the same site. From Saxon and Norman times onwards, there is generally some form of documentary evidence to explain either the foundation of a new settlement or the abandonment of an old one.

a) The prehistoric period

During the first half of the 10 000 year period between the end of the Ice Ages and the Roman invasion of Britain, evidence of settlement is very scanty and mainly takes the form of bones and artefacts that indicate where people lived (in locations such as caves and along shorelines), rather than what their settlements looked like. As these 5000 years saw great fluctuations in climate, settlement locations would have varied considerably in terms of altitude. The first period in which permanent farmsteads and hamlets were widespread throughout most of Britain was the Neolithic, which dates from around 3300 BC, when the culture of farming and pottery-making was introduced to southern England from France. The Neolithic, followed by the Bronze Age and then the Iron Age were three millennia in which gradual improvements were made in tool-making and farming practices, which enabled the colonisation of the land to become more widespread.

The main settlements to have been discovered from this period tend to be located on higher land with lighter soils, which could be readily worked by the simple ploughs that were used by the peoples of these early cultures.

In the south-west of England some of the biggest concentrations of Bronze Age settlements are to be found on the slopes of Dartmoor, at elevations between the 250 and 500 metre contours. These villages of circular huts and attached field systems would have been very productive in the warmer climate of around 1500 BC, but are on land that is today very marginal. One of the largest of these abandoned settlements, at Carn Brae, had over 60 huts. By the late Iron Age considerable regional variations in building styles had developed and some of the most complex villages were to be found in western Cornwall, where farmers lived in courtyard houses with plots of land attached to them. Chysauster is the best preserved of these excavated sites.

b) The Roman occupation

The Romans brought with them new technology and administrative systems that changed the face of the countryside. As they were able to work deeper soils of the valleys and lowlands, the colonisation of the land spread over the entire civil zone of their occupation. Rather than totally transforming the countryside, they added to it; native Iron Age village settlements would have continued side by side with Roman villas. The villa, which was both an estate farm and a country home for important town-dwelling Romans, was the most significant new element that the Romans, added to the British landscape. Some areas of the lowland zone of Britain are particularly rich in remains of Roman villas, such as the area around Ilchester in Somerset, the Cotswolds and the Hampshire basin.

There are hundreds of abandoned Roman villas throughout England and Wales. When the Empire was beginning to disintegrate in the early years of the fifth century AD and the armies were called back to defend the less peripheral parts of the Roman territories, what happened in Britain is still a matter of debate. Were so many villas abandoned immediately after the fall of Rome in 410 AD or did it happen much later when the skills needed to maintain and repair them were forgotten? Once again, there is the question of continuity. There are several sites that have been excavated, such as Withington in Gloucestershire and Cheddar in Somerset, where there is definite evidence that Roman villas continued to be occupied in Saxon times.

c) Saxon and Norman colonisation

The period of the Anglo-Saxon settlement of Britain from roughly the fifth to eleventh centuries was very much a period of clearing of forests and the colonisation of the lowlands. It was therefore very much a period of consolidation rather than abandonment. As Hoskins writes:

> The Anglo-Saxon settlement was spread over some twenty generations between about 450 and 1066. During this time England became a land of villages.

By the end of the Saxon period, the population was sufficiently settled and high in density for all villages and hamlets to have permanent names, and well-defined and documented boundaries, in order to prevent disputes over territory. There were in the north and east of the country Viking raids and the establishment of Viking communities. The arrival of the Normans was merely a take-over of the existing rural landscape, with a few new villages established and others just given Norman French additions to their names.

d) The Black Death

One of the biggest changes to the settlement pattern of Britain to take place in the Middle Ages resulted from the Black Death. The bubonic plague hit the country in the late summer of 1348. Although there are more than 2800 deserted villages in England, the vast majority of these desertions belong to a later date than the Black Death. The main effects of the plague were to partially depopulate villages, take away the pressure upon the land and so less favourable places were therefore abandoned.

e) The Late Mediaeval and Tudor sheep enclosures

The vast majority of the Deserted Mediaeval Villages (DMVs) in England were abandoned for economic reasons rather than being hit by the plague. During the late Middle Ages when there was peace throughout much of western Europe and England had close economic ties with Flanders, France and the states of northern and central Italy, English wool was highly prized and therefore the country's main export. Demand for wool led to the destruction of hundreds of villages and their communities throughout England, as landowners converted from the traditional open common field systems of cereal production to sheep farming.

The biggest concentrations of DMVs from this period are to be found in Midland England, in the counties of Warwickshire, Oxfordshire and Northamptonshire, as well as further north in Yorkshire (see Figure 5.2). Ordnance Survey maps of these areas provide the evidence for village desertion with the frequently occurring words 'site of village' and the mark of a cross. In the field, the evidence may

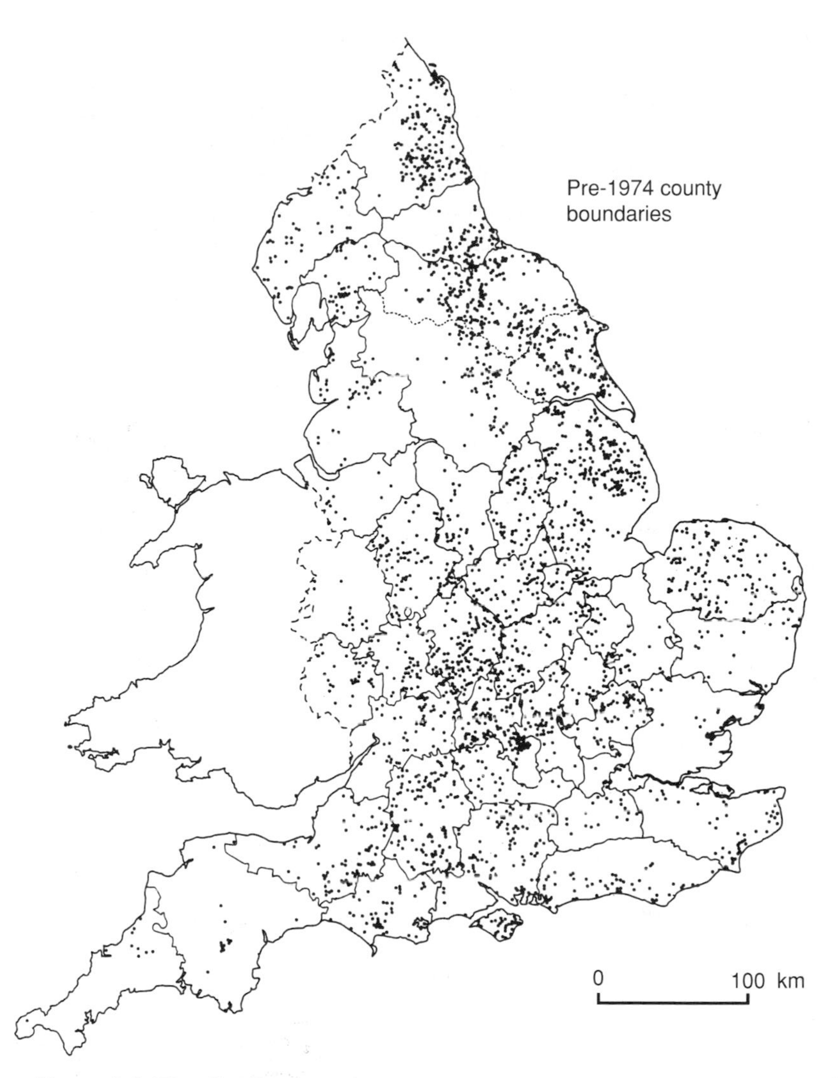

Figure 5.2 The distribution of Deserted Mediaeval Villages in England

be in the form of a few overgrown ruins, but more commonly the old ridge and furrow pattern of the open common arable fields can still be seen fossilised in the grass.

f) The Dissolution of the Monasteries

Although some of the monastic houses in Britain had been suppressed in earlier times, such as in the reign of Henry V for anti-French propaganda reasons, the systematic destruction of the monasteries took place from 1536 onwards following Henry VIII's break with Rome. Many of the rural monasteries were complex settlements supporting large numbers of laity as well as monks and would have been engaged in farming hundreds of hectares of land. Although many of the monasteries were already on the decline by the reign of Henry VIII, many of the larger ones such as Fountains Abbey in Yorkshire were thriving and supported hundreds of people, either directly or indirectly, in both farming and small-scale industries. With dissolution there came further waves of dispossessed rural populations searching for employment in the cities.

g) The seventeenth to the nineteenth century

The patterns set by the landowners in the Tudor period continued for three centuries. As each lord of the manor wished to outdo others, country houses and their attached parklands became grander. More traditional villages were eradicated from the landscape as they 'spoiled the view' of the landscape gardens. It was this fashion that was satirised in Oliver Goldsmith's poem *The Deserted Village*, which was based upon the destruction of Nuneham Courtney in Oxfordshire that underwent the 'emparkment' process in the 1760s. Other examples of villages destroyed for similar reasons in this period include Milton Abbas in Dorset and Houghton in Norfolk.

The biggest and most notorious example of rural depopulation in this period were the 'Highland Clearances' in Scotland. The Highlands of Scotland remained much more feudal than other parts of Britain well into the eighteenth century. Highlanders were shipped off as either indentured labour or slaves to the New World by their lairds in the mid-eighteenth century. As with fifteenth century England, whole village populations in the Highlands were evicted to make way for sheep farming. Throughout the eighteenth and nineteenth centuries there was a steady flow of poor, evicted migrants to both the USA and Canada, particularly to places such as Nova Scotia and the other Atlantic provinces, to the Lowlands of Scotland and to England.

It was only in 1976 that crofters were first allowed legally to own their land. In 1997, the Scottish Industry Minister introduced an initiative *Iomairt air an Oir* (Initiative at the Edge) that is to address the issues such as depopulation in Scotland's peripheral areas, and to halt the flow of migrants.

h) The twentieth century

There were fewer examples of village desertion in the twentieth century than in previous periods in history. This does not mean to say that there was not a decline in the population of many villages. Remote rural areas saw a decline in population for most of the century, whereas those close to large urban areas started to gain population as a result of urban expansion and then counterurbanisation from around the middle of the century. One type of settlement that did start to disappear from the landscape was the mining village. Even in the nineteenth century when certain mineral workings became exhausted, small mining settlements became abandoned as they lost out to the 'economies of scale' of the larger industrial cities.

One of the biggest desertions of mining villages came as a result of the planning of the local authorities in County Durham. In the 1960s pit villages of County Durham were put into four categories as to what their fate would be. Category A villages were to become key settlements, Category B and C villages were to remain as living communities but with fewer services and developments. Category D villages were to have no investment in them whatsoever; this would allow them to remain static, shrink in size or even become no longer viable as communities, and eventually fall into abandonment and be demolished.

The full range of reasons for abandonment can be added to with a few more examples from the twentieth century. Continued urbanisation had increased the demands upon water supplies and throughout Britain new reservoirs were built, some drowning valleys with villages located in them, e.g. West End near Harrogate, that disappeared when the Thruscross Reservoir was built in the 1960s in Nidderdale. Several villages have been depopulated for military reasons. Imber, a village high on Salisbury Plain, was already suffering badly from depopulation in the 1930s, and in 1943 it was taken over by the army in order to train soldiers for street fighting. With its church, pub, manor house and other houses semi-boarded up, Imber is still used for army practice today.

3 Rural Settlement Abandonment Elsewhere

CASE STUDY: SETTLEMENT ABANDONMENT IN BELARUS

Government policies can have a profound effect upon the success and continuation of rural settlements. In the former Soviet Union, the policies of collectivisation of agriculture from the 1930s onwards are still having an impact upon the republics that once made up the USSR. Many of the countries of eastern Europe have negative population growth rates and some of the most rapidly ageing populations in the world. At the same time, in the attempts to make the transition to a market economy, these countries are undergoing severe economic and social problems that are often felt more strongly in the countryside where income levels are very much lower than in the cities. Belarus, the Baltic Republics, the Ukraine and the Russian Federation itself all fall into this category of having serious problems of adaptation to change in rural areas.

A Human Development Report from the UNDP (United Nations Development Programme) in 1996 identified that Belarus has over 630 abandoned villages dating from the post-Communist period. The biggest numbers of abandoned settlements are found in the Vitebsk and Gomel oblasts (regions), located close to the border with Russia and the smallest number in the oblast administered from Minsk, the centrally placed capital. This would suggest that in Belarus there is a distinctive core versus periphery effect within the levels of prosperity.

Until the changes initiated in the 1930s by Stalin, Belarus had a well-connected network of farms, hamlets and small villages. During the forced collectivisation of agriculture about 190 000 farms and villages were either destroyed or merged into bigger units. During World War II many villages were destroyed or badly damaged by the Nazis. In the decades following 1945 there was heavy rural–urban migration as the cities of Belarus expanded their heavy industries, creating big demands for additional urban labour force.

At the same time the post-war reconstruction of farms and villages was done on the cheap, leading to sub-standard housing conditions. There was further merging of collectives to make them into larger units. The government ruled that the average collective farm size should be between 1000 and 1500 hectares and worked by between eight and 15 villages, depending on their population size. In the 1960s the average collective farm size was pushed up to between 2500 and 4000 hectares. One of the draconian measures introduced to achieve this expansion of farm size was to identify villages as being either 'promising' or 'unpromising' enabling the government to manipulate the overall settlement structure and hierarchy. The 'promising' villages were to be given extra amenities, whereas the 'unpromising' ones were to be depopulated. A total of 5800 villages were selected for 'promising' status, whereas 8400 were allotted 'unpromising' status; of the latter only 1500 were forcibly depopulated with their people moved to larger units, the rest were left to disappear 'naturally' as their populations aged and emigrated.

The main reasons for the continued decline of rural settlements in Belarus include:

- the ageing population and therefore workforce
- the declining quality of the physical environment, soil fertility and natural resources
- the declining quality of rural housing and lack of funds to make improvements
- the decline in the number of services provided in rural areas
- the poor levels of amenity provision, e.g. piped water and gas supplies.

CASE STUDY: CONFLICT AND RESETTLEMENT IN RURAL RWANDA

Few countries have had such a violent recent history that has displaced such large percentages of the rural population as Rwanda. This small east African state with a population of 7.3 million people is dominated by the struggles between its two main ethnic groups the majority Hutus (80% of the population) and the Tutsi minority (15% of the population), who traditionally formed the ruling class. Each of these two groups has had the dominant hand in running the country, which has led to the other group being expelled in large numbers and returning later as refugees.

Land ownership and the types of rural communities have made the situation a complex one. Traditionally the Rwandans saw land as a natural resource rather than private property. In the seventeenth century the power of the ruler, the *umwami*, increased and he became the owner of all the land but granted people its use. With the connivance of the European colonisers the ruler tried to charge farmers rents for the land; this worked in some regions but in others kin groups retained rights of ownership over the land they cleared, known as *ubukonde*. With independence most land came into the hands of the state and the Hutus and Tutsis came into conflict over it.

In 1962, just before independence, the Hutus overthrew from power the Tutsis, which led to a refugee population of some 120 000 leaving the country. In the early to mid-1990s the Tutsi refugees formed an army and invaded, destroying large areas of Hutu land and causing around 2 million of them to flee to neighbouring countries, especially Zaire and Uganda, where living in appalling conditions in refugee camps they were still at the mercy of Tutsi gunmen. In both periods of violence there was wanton destruction of people, villages, crops and animals, making Rwanda one of the twentieth century's worst centres of humanitarian crises.

Since the Arusha Accord peace treaty of 1996, the Hutus have been returning to Rwanda and being resettled upon the land. One of the biggest problems to be solved is that of rights over the same lands being claimed by both ethnic groups, causing continued friction between them.

The Rwandan government in 1997 established an extensive land and settlement reform programme throughout the contested areas of the country. A 'grouped' rural settlement type known as *imidugudu* was established as the optimum way of developing the rural sector and restructuring after the civil war. These grouped villages would enable framers to work together on areas of up to 50 hectares as opposed to many of them struggling to make a living from tiny plots of land in the past. The land reform has deliberately distanced the farmers from their land in order to break their emotional attachment to it, which had caused so much conflict in the past. Although many successful *imidugudu* have been established close to Kigali, the capital, progress is much slower in the more remote regions where less money has been invested and some of the old rivalries continue.

4 Rural–Urban Migration and Rural Depopulation

One of the most important forces behind rural depopulation in the modern, industrialised world is rural–urban migration. No doubt this type of migration had plenty of antecedents in the pre-industrial world when people from rural areas sought their fortunes in the big cities of the ancient world or of Mediaeval Europe. Also for millennia, there would have been countless examples of displaced rural populations fleeing the countryside because of problems such as famine and war and taking refuge in the comparative safety of their nearest town or city.

It was not until the Industrial Revolution in Britain that rural–urban migrations began to take place upon an unprecedented scale. The rural depopulation that Britain experienced in the century between 1750 and 1850 was due to two groups of changes: those taking place in the cities and those taking place in the countryside. In the cities, the rapid expansion of industries based upon the factory system was creating huge new demands in labour supply; this led to very rapid urbanisation and the swallowing up of many farms and villages that had been on the edges of small towns, such as Manchester, Bradford and Sheffield, making them part of the urban fabric.

Similarly, large numbers of people were migrating to the newly developing coalfields and other mining areas that transformed what were previously essentially rural areas such as the valleys of South Wales into urban ones. The displacements of population from rural areas to urban ones took place at the local, regional and national levels, and the development of the railways from the 1830s onwards enabled long distance migrations to become more common. At the same time as urbanisation was taking place in Britain, the Acts of Enclosure were going through Parliament. If the new industries in the cities provided the pull factors of migration the Enclosure Acts created an important push factor. The Enclosure Acts led to the end of the old system of subsistence agriculture in Britain and were one of the measures that led to the commercialisation of farming. Along with the new technologies of the Georgian Age, such as selective breeding, farm machinery and new crops and forms of crop rotation, Parliamentary Enclosure led to a great leap forward in productivity on the land. It also led to a sharp decline in the number of people involved in working the land. The open common arable fields that had dominated a village's supply of grain ever since Saxon and Norman times disappeared, as did much of the common grazing land on hilltops and valley bottoms. The newly enclosed common lands were consolidated into just a few farms per village, putting hundreds of thousands of villagers out of work, and these became the bulk of the migrants to the cities. With these changes in the countryside, individual farmers may have prospered, but villages themselves often went into decline through population loss. In particular it triggered off the problems of the ageing of rural society, with the emigration of able-bodied workers; many villages never properly recovered from this. The situation was merely exacerbated in the second half of the nineteenth century with the repeal of the Corn Laws, which protected Britain's farmers from foreign competition, a series of disastrous grain harvests at the end of the century (which may have been due to temporary climate change caused by the Krakatoa eruption in Indonesia) and the flooding of the markets with cheaper grain from abroad especially from North America.

In Britain the rural–urban migration and the loss of rural population continued down until the second half of the twentieth century when population trends started to reverse. Elsewhere in Europe it has taken place over different time scales. In the more peripheral and marginal areas of the EU rural depopulation continues today, e.g. central France, southern and insular Italy, the interior of Spain, northern Sweden, the poorer lands of the old Democratic Republic of (Eastern) Germany. Such areas suffer from all of the problems of decline, which follow on from population loss, including an ageing population, the decline in rural services, rural poverty and rural isolation

The most important streams of rural–urban migration today are taking place in LEDCs. Many of the trends are more extreme than in Europe in the nineteenth century because of better transport and communications. Furthermore, the patterns of rural–urban migration seem to take a different form: **step migration** seems to be common in LEDCs, and this often leads to the rapid and unbalanced growth of one major city (generally the capital) and its region, to the detriment of other parts of the country. Step migration involves people first moving from village to small town and then working up through the hierarchy, whereby a disproportionate number of people end up in the capital or at least the largest port if it is not the capital.

CASE STUDY: THE RELATIONSHIP BETWEEN ALTITUDE AND DEPOPULATION IN THE ABRUZZO REGION, ITALY

Most of the Italian regions have suffered from rural depopulation for over 100 years. The big outward migrations started in the last two decades of the nineteenth century when a sequence of bad harvests created desperate conditions in what were already overpopulated rural areas. Higher altitude settlements generally have more marginal land than lower altitude villages and one of the best ways of understanding the impact of rural–urban migration is to compare current average heights at which the rural population lives with those in the past. The following are examples of the altitudinal changes of the average rural population, experienced between 1921 and 1991 for selected Italian regions (heights in metres):

	1921	1991	% change
Liguria	87 m	55 m	–36
Emilia-Romagna	138 m	84 m	–61
Lazio	254 m	127 m	–49
Abruzzo	486 m	359 m	–28
Basilicata	614 m	572 m	–7
Sicily	296 m	223 m	–27

The Abruzzo is one of the poorest of the Italian regions and has suffered heavy outward migration from its rural areas. The region is predominantly rural and stretches from the Adriatic coast in the east, westwards through highly dissected and eroded hill country up to the highest limestone peaks in the Apennines, several of which reach over 2500 metres (see Figure 5.3). Apart from the coastal plain where the port city of Pescara is located, the other most suitable places for settlement are the intermontane basins of the Apennines where the Abruzzese capital, L'Aquila, and other major centres of Avezzano and Sulmona are located. The other two most important towns of the region, Chieti and Teramo, are positioned in the foothills of the Apennines, just above the coastal plain. All these cities occupy key positions that have contributed to their prosperity and importance, through both their rich agricultural hinterlands and their domination of routeways. The vast majority of rural settlements in the Abruzzo have not been as fortunate in their locations and levels of prosperity.

The Abruzzo, like all other parts of central Italy, has had a very turbulent history and the mountains are scattered with hundreds of abandoned villages and hamlets. There are two main reasons for abandonment in the past: natural causes, especially earthquakes (the last major one struck in 1915 affecting the Avezzano area), and human destruction from Adriatic invaders or civil strife. The past abandonment of rural settlements has been paralleled by a more recent, gradual process of depopulation resulting from economic migrations. So many of the villages in the Abruzzo are perched in high mountain-top or mountain-side positions

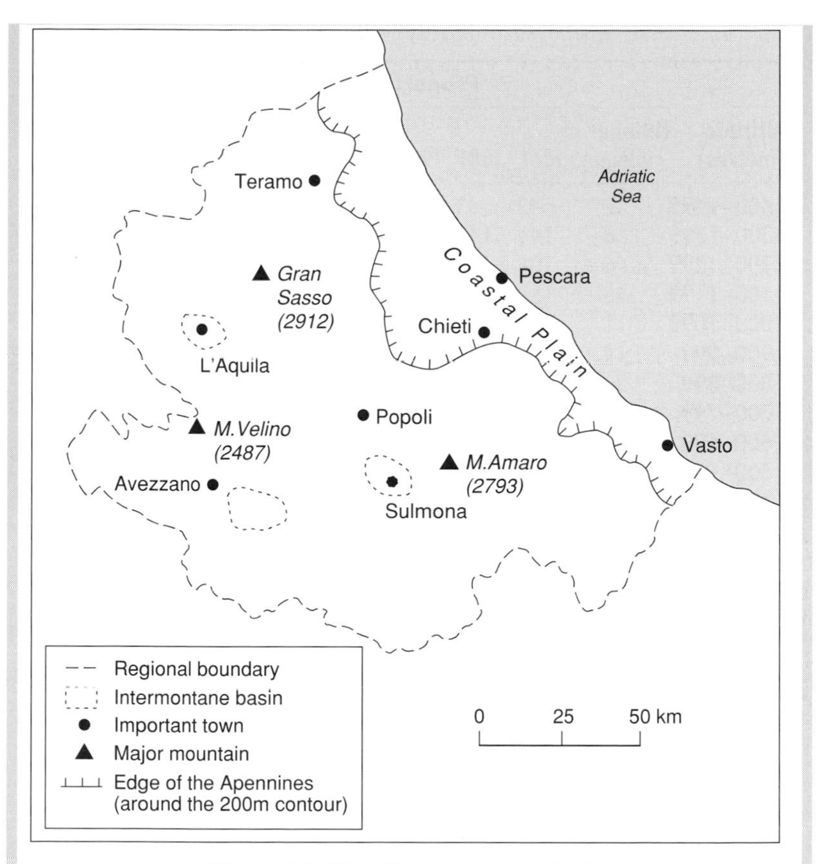

Figure 5.3 The Abruzzo region, Italy

because of the past need for defence from invaders. This has left them at some distance from their agricultural land, which has made the daily journey to work arduous, especially before wide ownership of a basic motor vehicle. Remoteness from towns and cities, severe winters, hot dry summers, poor diet, land tenure and low wages presented problems to the local inhabitants, leading to heavy emigration from the mid-nineteenth century onwards, particularly accelerated by the bad harvests of the 1890s.

Although there was some slowing down of emigration in the 1920s because of the actions of the Fascist Regime, the post-war period saw particularly high outward migration associated with the economic boom in the industrial cities of the north of Italy. In some parts of Italy this rural depopulation has been reversed (e.g. counterurbanisation has added to the population of the villages of the Roman Campagna in Lazio). In mountainous areas

Table 5.1 Chart of population changes in Abruzzo

		Proportion of population 0/000								
Altitude (metres)	**Number of villages**	**1861**	**1881**	**1901**	**1921**	**1936**	**1951**	**1971**	**1981**	**1991**
1400–1499	2	42	43	37	40	35	31	20	13	11
1300–1399	6	144	145	142	137	110	97	68	55	48
1200–1299	6	108	106	94	72	60	54	44	39	35
1100–1199	6	152	150	138	128	116	103	79	61	51
1000–1099	11	234	241	226	201	190	180	134	113	98
900–999	14	381	369	357	338	284	251	163	135	123
800–899	30	594	617	603	592	528	474	351	311	298
700–799	27	1133	1180	1131	1084	1026	991	969	932	925
600–699	28	854	864	873	842	823	794	706	672	671
500–599	24	615	595	586	588	550	516	367	329	303
400–499	45	1785	1733	1700	1680	1661	1647	1500	1463	1439
300–399	34	1276	1277	1323	1346	1338	1370	1267	1234	1221
200–299	42	1453	1431	1452	1465	1516	1515	1392	1393	1421
100–199	19	675	678	692	703	750	754	812	894	960
0–99	11	557	570	645	784	1014	1222	2128	2357	2396

of the Abruzzo region, however, there is a direct correlation between altitude and rural depopulation (see Table 5.1).

It can be seen clearly from this chart that all *comuni* located over 800 metres have continuously suffered population loss in the last 140 years, and this is still happening today. These villages are those with the most marginal farmland and the poorest road links with the outside world. *Comuni* located between 300 and 800 metres have shown only a moderate decline in their populations. Villages between 200 and 300 metres are in a critical location, as they have had the greatest fluctuations population levels over the 140 year period. By contrast all the villages lying between 100 and 200 metres have shown a steady increase in population, whereas those located below 100 metres have had a very heavy increase. The development of the coastline of the Abruzzo region as a string of holiday resorts is one of the main reasons for this latter situation.

Some of the villages located above 1200 metres are not losing population at the same rate as others. These *comuni* are stabilising their populations because of tourism; the villages located here have access to mountainsides located at altitudes of up to 2000 metres and therefore a potential skiing season that lasts for 4–5 months. In villages such as Roccaraso, Rivisóndoli and Ovíndoli, the agricultural economy has therefore been widely supplemented by winter sports, even though this is only a seasonal activity.

One aspect of rural depopulation is how much it influences the population structures of the villages it affects. It is generally the young that move away in search of work, and more men than women tend to migrate for this purpose. This can leave whole communities debilitated and set them on the road to a spiral of economic decline. In the higher altitude villages of the Abruzzo, as elsewhere in central Italy, where winter sports have no potential, an ageing population is one of the greatest problems.

Figure 5.4 helps to illustrate the degree of ageing in the population of rural Abruzzo. The first of the three population pyramids is that of the whole Abruzzo region and therefore includes urban as well as rural settlements. The graph shows that the two lowest age groups are contracting, partly due to continued migration away from the region and partly due to the decline of the birth rate within Italy as a whole. The second pyramid is that of the village of San Salvo, which is located at an altitude of 101 metres above sea level and is not far from the Adriatic coast; this places it in a reasonably accessible position and therefore it is ageing at a slightly slower rate than the average for the Abruzzo region. By contrast, the third graph is for the mountain village of Villa Santa Lucia degli Abruzzi, which is located at 850 metres above sea level and is consequently still losing people through outward migration and experiencing severe problems of an ageing population.

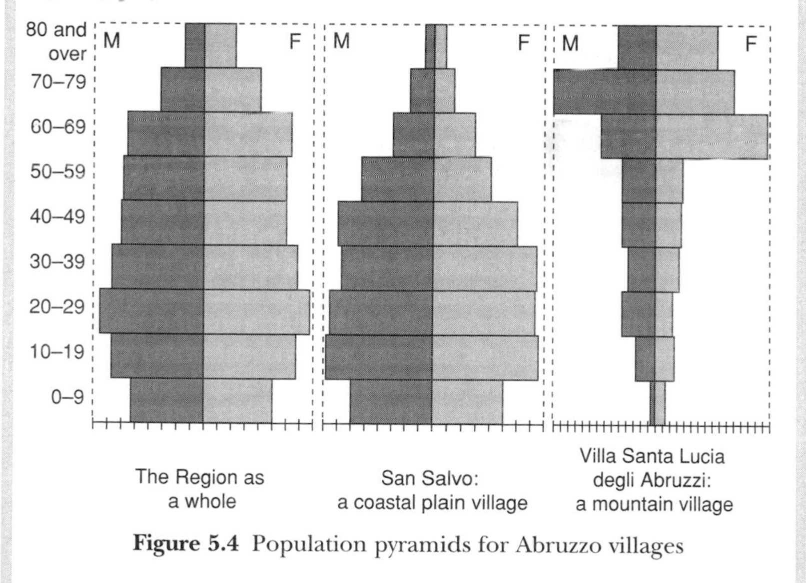

Figure 5.4 Population pyramids for Abruzzo villages

5 The Effects of Urbanisation on the Countryside

Both the outward growth of cities and the ways in which cities come to dominate whole nations demographically have a huge impact upon the countryside. The biggest impacts are naturally upon the rural areas immediately surrounding rapidly expanding large cities. Urban sprawl is the biggest threat that cities impose upon the countryside. More land is required for the outward expansion of housing and other urban functions. The sprawl is most pronounced along major roads where it takes the form of **radial development** or **ribbon development**.

The outward expansion of cities and the consequent sprawl have created a whole series of inter-related problems, which include:

- the loss of farmland and other open spaces
- the destruction of natural habitats, biodiversity and the threatening of rare species
- increased pollution leading to a decline in the quality of the air and water
- new roads need to be built to ensure connectivity of newly built suburbs eating up more green space
- and creating greater access of a larger number of people to a shrinking countryside
- an ever increasing pressure upon services and public utilities
- greater demands upon the surrounding countryside for leisure facilities, e.g. golf courses, and public utilities, e.g. rubbish disposal sites, new reservoirs for urban water supply
- the loss of 'tranquil areas' (as defined in Chapter 1)
- a greater built environment can have a knock-on effect upon the local physical geography, e.g. flashier streams and rivers causing more frequent flooding; more extreme heat island effects.

6 Urban Sprawl and the Role of Green Belts

The biggest effects that urbanisation have on the surrounding countryside are the problems of outward growth, urban sprawl and the lack of its containment. Land use planning around the edges of large urban areas is critical for the containment of urban sprawl and the most effective way in which this is achieved is through the creation of **green belts**.

Green belts have their origins in the writings and actions of the nineteenth-century fathers of modern town planning such as Ebenezer Howard who developed the concept of Garden Cities and the philanthropic industrialists such as Robert Owen, Titus Salt, the Cadbury family and the Lever family. The Town and Country Planning Act of 1947 led to the setting up of Britain's first green belt around London in the same year. There are now 15 well-established green belts around English metropolitan areas and conurbations covering about 4.5 million hectares. The largest of these is London's green belt, which covers 1.2 million hectares.

Green belts have been a qualified success in Britain, and the London green belt has been a good example of both the successes and failures of the concept. On the positive side London has benefited in the following ways:

- the green belt has protected a great deal of farmland
- the distinction between London itself and the towns beyond the green belt has been retained
- the belt has had some effect in limiting commuter flows because of that distinction
- more recreational land has been preserved than if the green belt had not existed.

On the negative side too many breaches have been made in London's green belt, these have limited its effectiveness:

- most of the M25 ring road has been built upon the green belt and it has degraded areas of environmental importance, e.g. its close proximity to the North Downs Way between Caterham, Surrey and Westerham, Kent
- there is so much pressure upon greenfield sites for housing developments; Britain needs over 4 million new houses by 2016 and about 25% of these will be built in the south-east of England
- social and economic considerations may outweigh environmental ones, e.g. the building of new hospitals or high-tech research facilities may take priority over conservation
- the management of green belt land is often poor, leaving it unattractive to locals and city dwellers alike; some parts have even become 'Edgelands' (see page 114)
- the green belt has encouraged more developments on greenfield sites around towns set in the countryside beyond the area it protects, e.g. around Milton Keynes, Stanstead in Essex and Ashford in Kent.

In contrast to London's well-established green belt is the relatively new ***Ruota Verde*** ('Green Wheel') plan for the city of Rome first put forward in the mid-1990s by the Green Party mayor, Francesco Rutelli, and adopted as part of the city's **structure plan** in the year 2000. The green areas take up an area of 82 000 hectares; although this is very small in comparison with London's green belt, it must be borne in mind that Rome is a much smaller and more compact city. The 'Green Wheel' contains two elements:

- a green belt (**cintura verde**) made up form farmland, large areas of forest, *macchia* (scrubland) and regional nature reserves
- green wedges (**cunei verdi**), formed by 'archaeological parks' (e.g. the Appian Way), urban parkland (e.g. the Villa Boghese) and a series of newly formed urban nature reserves.

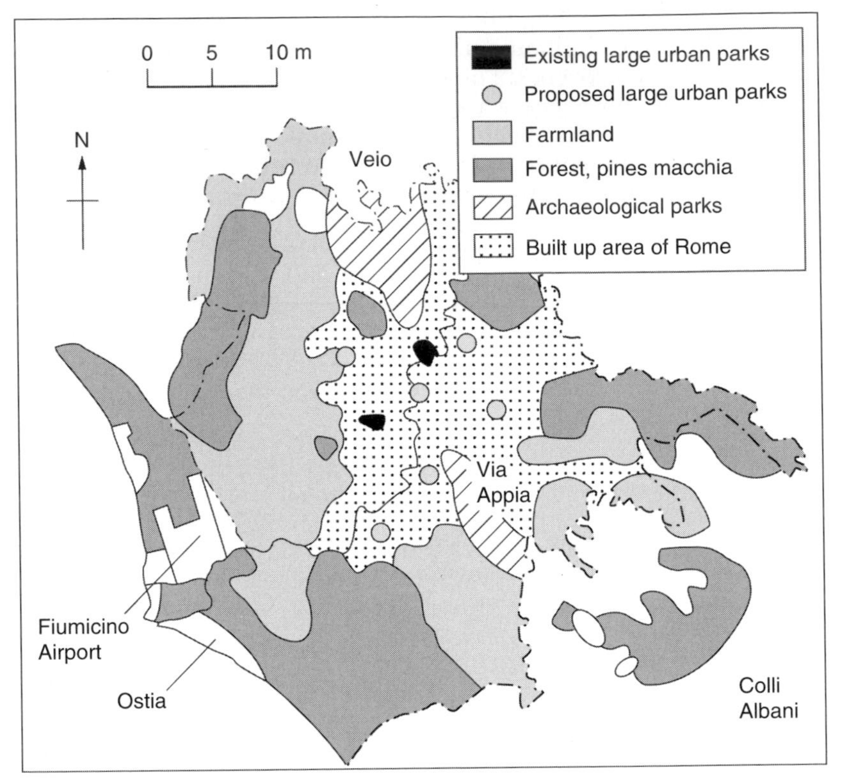

Figure 5.5 Rome's green belt

Figure 5.5 shows the main aspects of the Roman 'Green Wheel' plan. Although the plan is both comprehensive and well devised, it needs to be properly enforced to be effective. The main threats to the 'Green Wheel' could be:

- the development of any large-scale infrastructure schemes may eat into the green areas, and these might be of national importance
- as in London, a prestige development of social or economic importance could outweigh the environmental costs
- complex planning laws have often led to the building of illegal houses on the city's rural fringe in the past, and this might be likely to continue into the future
- political corruption has led to planning infringements in the past and could continue in the future.

CASE STUDY: THE ABSORBTION OF FARMS INTO THE URBAN FABRIC OF THE CITIES OF MILAN AND LECCE IN ITALY

As cities grow outwards they engulf both farms and farmlands. The valuable farmland of the city fringe – typically used for intensive activities such as market gardening, chicken or pig farming – then becomes even more valuable as land for housing or some other form of urban development. Actual farm buildings may be demolished, or if they are of historic value may be converted into luxury dwellings. Farm buildings can end up by being converted for numerous different uses once they become absorbed into the fabric of the outward expanding city.

The research carried out by two Italian geographers, Grosso working in Milan and Quarta in Lecce, provide two contrasting studies of how old farms have been used within the cities that have engulfed them. In both cases, the traditional farms are large-scale complexes of houses and outbuildings, and therefore most have been saved and adapted rather than being demolished. This is where the similarities end and the economic, social and political differences between northern and southern Italy come into sharp contrast. Figure 5.6 shows the distribution of the farms, their state of repair and their functions. In the case of Milan, a city of some 1.3 million people, there were 45 *cascine* (the general term used for estate farms in the Lombardy region) within the urban confines surveyed in Grosso's study. Of these, 13 were found to be empty or occupied by squatters. Thirteen of the farms still had some kind of agricultural use (e.g. urban farms, food wholesaling or storage) and a further two had been converted to other commercial use. Only five of the farms had been converted to residential use, either as flats or luxury homes. The most striking aspect of the new uses given to the old farms has been in the public sector. Twelve of the properties have been converted to public, social and cultural functions; these range from art galleries and studios through to public offices and centres for old people. The situation in Milan shows a high level of local government intervention and public benefit from the rehabilitation of the old *cascine.* This case study also reflects the comparative wealth of the city and people of Milan in being able to carry out these schemes.

By contrast, in Lecce the situation is very different. With just under 100 000 people, the city is much smaller than Milan and is set within a much less prosperous part of the country, which still

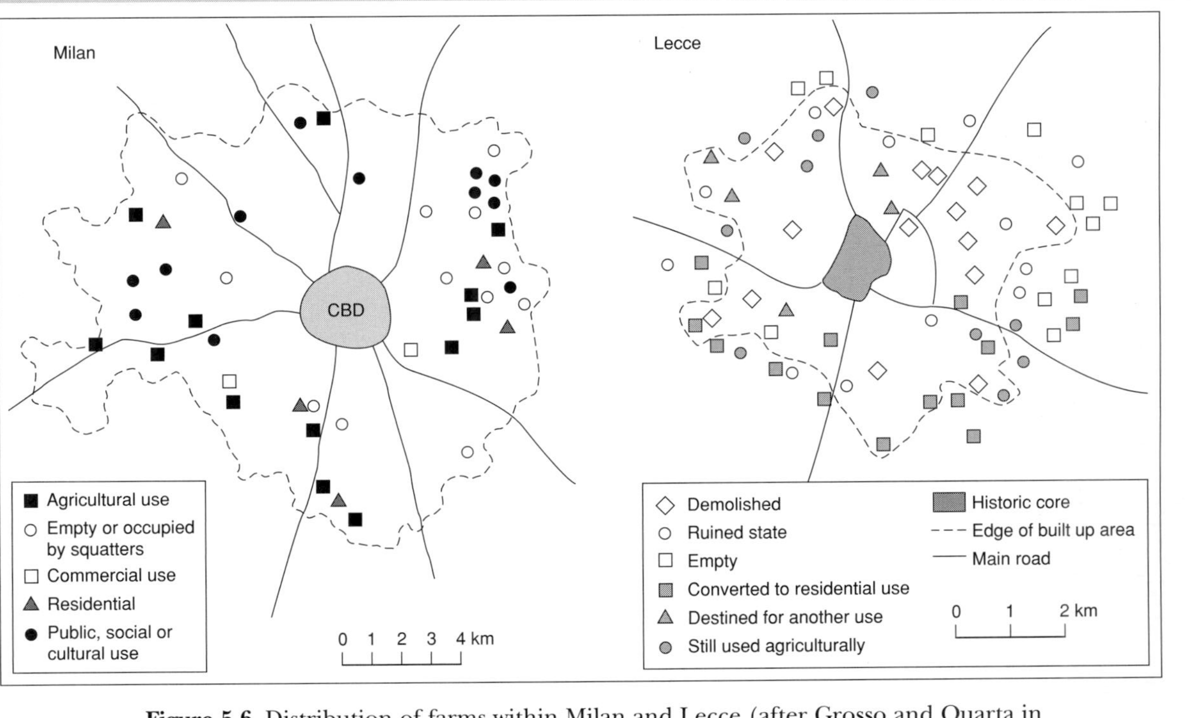

Figure 5.6 Distribution of farms within Milan and Lecce (after Grosso and Quarta in *Bollettino della Società Geografica Italiana*, 1996)

has a high proportion of people working in agriculture (13% of the workforce in the region of Puglia as opposed to just 2.8% in Lombardy). Of the 68 *masserie* (the name widely used for estate farms in this part of Italy) studied by Quarta, 39 were either being demolished, derelict or empty. Ten of the farms were still found to have an agricultural use, mainly farms still operating on the city edges. Fifteen farms have been converted to residential use, some of these have produced rather substandard speculative housing and only six farms are destined for other uses. This all gives a very different profile from the farms that have been absorbed into the urban fabric of Milan. In Lecce the shortage of money, the lack of political will and the length of time it takes to buy and sell property are all reasons why the farms have not been put to better use.

CASE STUDY: THE IMPACT OF THE GROWTH OF DELHI ON ITS RURAL HINTERLAND

Delhi, India's capital, provides a very interesting and typical example of what is happening to the countryside surrounding large cities in LEDCs. With a population of over 10 million, Delhi is one of India's megacities. Located on the densely populated Indo-Gangetic Plain of northern India, it has a relatively flat site that has led to urban sprawl in all directions; however, it is growing more rapidly to the north than to the south, where the low Aravalli Hills provide a natural barrier that is limiting urban development.

This more attractive southern fringe of Delhi has been where the rich and the middle classes have built their country villas. On this side of Delhi are also some satellite towns where both transnational and Indian manufacturing firms have located modern factories. By contrast, the area to the north of the city is rapidly developing the sort of functions that enable the growth of Delhi, with small-scale industries such as brick kilns, rice mills and other food processing plants, as well as storage facilities for the city.

J.V. Bentinck, the Dutch geographer carried out a detailed study of Delhi's hinterland and examined the changes that are taking place in selected villages on its urban fringe. The study concentrates on one of the four 'Blocks' or regions surrounding the city, the Alipur Block. This district, which is some 220 km^2 in area, lies to the north of the city and is bisected by the Grand Trunk Road, the main artery leading north west to Amritsar.

Using census data Bentinck identified four distinct stages in the change of villages from rural to urban characteristics:

- stage 1 'rural': both the workforce engaged in agriculture and the amount of land under cultivation both stand at over 50%
- stage 2 'occupational change': the amount of land under cultivation is still over 50%, but the workforce in farming has dropped below 50%
- stage 3: 'increasing urban land use': between 20% and 50% of the land is left under cultivation
- stage 4: 'urban': less than 20% of land is still under cultivation.

In the last ten years an increasing number of villages in the Alipur block have moved up through the various stages of the model. Those in stage 4 are not surprisingly those close to the city edge or along the Grand Trunk Road; those that remain in stage 1 are located along the flood plain of the Yamuna River where crop yields are highest and urbanisation is deterred by the hazard of flooding.

Khushk village, located close to the Yamuna River, is still in the 'rural' stage and its population is predominantly of the Saini caste who specialise in horticulture. Intensive vegetable and fruit production on the rich alluvial soils and aimed at Delhi's markets have kept agricultural employment at high levels. Furthermore, the Saini are proud of their traditional agricultural lifestyle and are loath to give it up. Some land is rented out for brick manufacture but small-scale industrial employment is served mainly by the very limited number of landless lower caste workers in the village.

Jagatpur is an example of a village in the 'occupational change' stage. Although located close to the city, it lacks good transport links. 80% of the population is of the Gujjar cattle rearing caste and therefore dairying is the village's agricultural specialism. As dairying is not as labour intensive as horticulture and because there are severe risks of flooding within the river area of the village lands, an increasing number of people are being employed in sectors other than farming. Jobs in the government services, commerce and shops selling household goods and building materials are attractive alternatives to agriculture. Many households combine incomes from owning dairy cattle and working in other sectors.

Nangli Poona is a village in the 'increasing urban land use' stage. Located close to Delhi it has had a long history of urbanising influences; in the 1930s when a radio station was set up there, the government acquired 20% of the village land. About 40% of the land is still under agriculture, with wheat, rice, vegetables and fodder crops being grown at different times of the year. Industry has spread along the GTR to Nangli Poona and the village now has 32% of its land taken up by manufacturing, infrastructure and commercial activities, giving it a semi-urbanised appearance.

Samaipur has become dominated by its proximity to Delhi and is in the 'urban' stage of development. Its population is much more mixed, both ethnically and in castes, than the other villages mentioned because of the large number of migrants and squatters. The village also has very mixed land use, which includes both traditional and new industries, patches still under cultivation and large areas of polluted, derelict land and sites cleared for new developments. The high-rise blocks of the city fringe loom up in the distance. Crime and corruption are on the increase and indeed appear to be closely tied in with the urbanisation process. Land grabbing (the making of false claims about the ownership of former farmland) is being stimulated by rising land values; at the same time there is considerable speculative development of 'farmhouses' – the euphemistic name given to the suburban villas of the city's wealthy middle classes. Sanitation is a huge problem as the population is expanding much more rapidly then the infrastructure. Perhaps the most striking feature of the chaos brought about by this particular stage in Samaipur's development is the growth of illegal economic activities. In late December 2002 the furnace of an illegal glass factory exploded killing nine workers and injuring numerous others.

As *The Hindu* newspaper put it:

> A visit to the area near Samaipur and the adjoining nullas (watercourses) and drains reveals startling facts about how unscrupulous factory owners and the connivance of corrupt babus (local government officials) are thriving and posing a big threat to the health of citizens.

Questions

1. Examine the various reasons why rural depopulation takes place.
2. Why do some rural settlements become abandoned?
3. What are the main types of impact that urban areas have upon their surrounding countryside and what planning methods are used to contain it?
4. Compare and contrast how the rural areas around MEDC cities and LEDC cities are affected by urban growth.
5. What factors cause the wide differences in rural depopulation rates between countries and between regions within them?

Summary Diagram

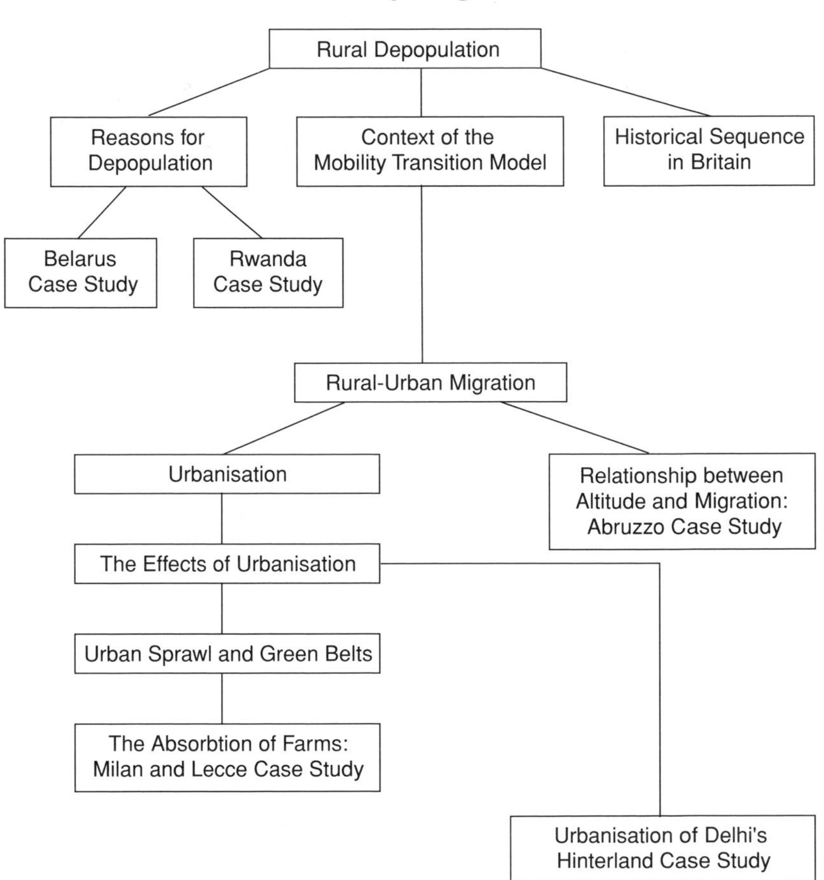

6 Counterurbanisation, Suburbanisation and the 'Post-Rural'

KEY WORDS

(Demographic) turnaround: the stage in a country's development when rural depopulation is replaced by counterurbanisation.
Counterurbanisation: the movement of people away from cities to the countryside and smaller settlements.
Re-urbanisation: the movement of people back to city centres (from suburbs or rural areas).
Rural business park: a planned zone of small factory/office units in a rural area.
Exurbs: a US term for a commuter belt area located between 10 and 50 miles from a large city.
Suburbanisation: the series of changes that take place on the edges of cities, their commuters belts and the countryside beyond as a result of counterurbanisation.
Gentrification: the upgrading of the housing of an area due to the in-migration of more wealthy people.
Post-rural: the adjective used to describe the countryside in which agriculture has ceased to be the main economic activity.

1 As one who long in populous city pent,
Where houses thick and sewers annoy the air
Forth issuing on a summer's morn to breathe
Among the pleasant villages and farms
5 Adjoin'd, from each thing met conceives delight

John Milton *Paradise Lost*

1 Demographic Turnaround and the Process of Counterurbanisation

Demographic turnaround is achieved when rural depopulation through rural–urban migration ceases to be a major trend within a country and it is replaced by **counterurbanisation**, which involves flows of migrants away from the larger towns and cities towards more rural areas. Counterubanisation can take place while some forms of rural depopulation continue. For example in Britain in the 1970s while cities were losing population to their surrounding areas, some of the remoter rural areas such as Mid Wales and the Outer Hebrides were still losing population to towns. Likewise in Italy today, the remoter villages of such regions as Sardinia and Basilicata are still experiencing rural depopulation whilst all of the major cities are undergoing counterurbanisation.

Counterurbanisation is a form of **step migration** both in the spatial and the historical terms. Spatially, counterubanisation can be likened to a 'ripple effect' in a pond. There are waves of migration going from the central core to the inner suburbs, from the inner suburbs to the outer suburbs and from the outer suburbs to the areas beyond the urban fringe. Of course, the migratory patterns are actually more complex than this, with for example movements from centre to outer suburbs and from inner suburbs to the rural fringe, but the basic concept of the ripple effect is a sound one. The historical sequence of the movements associated with counterubanisation has more to do with the increase of commuting distances and the discovery of ever remoter areas. The first waves of counterurbanisation were those affecting a city's outer suburbs, these would be followed by movements to the surrounding countryside, its villages and market towns, then the scope of the migrations would widen until eventually some migrants would move to the remoter parts of the countryside well beyond the field of influence of their city of origin. Figure 6.1 illustrates the role of demographic turnaround in relation to rural depopulation, counterurbanisation and suburbanisation.

2 The Spread of Counterurbanisation

As a result of turnaround, a city will pass from gaining population through rural–urban migration to losing it through urban–rural migration. Although the terms 'turnaround' and 'counterurbanisation' were first widely used in the 1970s when the phenomena were recognised and written about, the process of counterurbanisation had its roots in earlier decades. The first cities to start losing population from their main built-up areas to the suburbanised villages beyond their city limits were in the United States as early as the 1930s. It was, above all else, the mass ownership of motor cars that enabled a significant proportion of the wealthier inhabitants of US cities to live in, and commute from, places beyond the city limits. From the mid-1930s onwards commuting distances increased as more people sought a more rural lifestyle rather than that of the city proper. During these decades the overall populations of such cities as Los Angeles and Chicago were still growing rapidly, but so too were the populations of their surrounding rural commuter belts, often at much more rapid rates.

Counterurbanisation in Britain has its roots in World War II. London was the first British city to experience population loss in modern times; it reached its peak population of 8.6 million on the eve of World War II in 1939 (Inner London's population had peaked

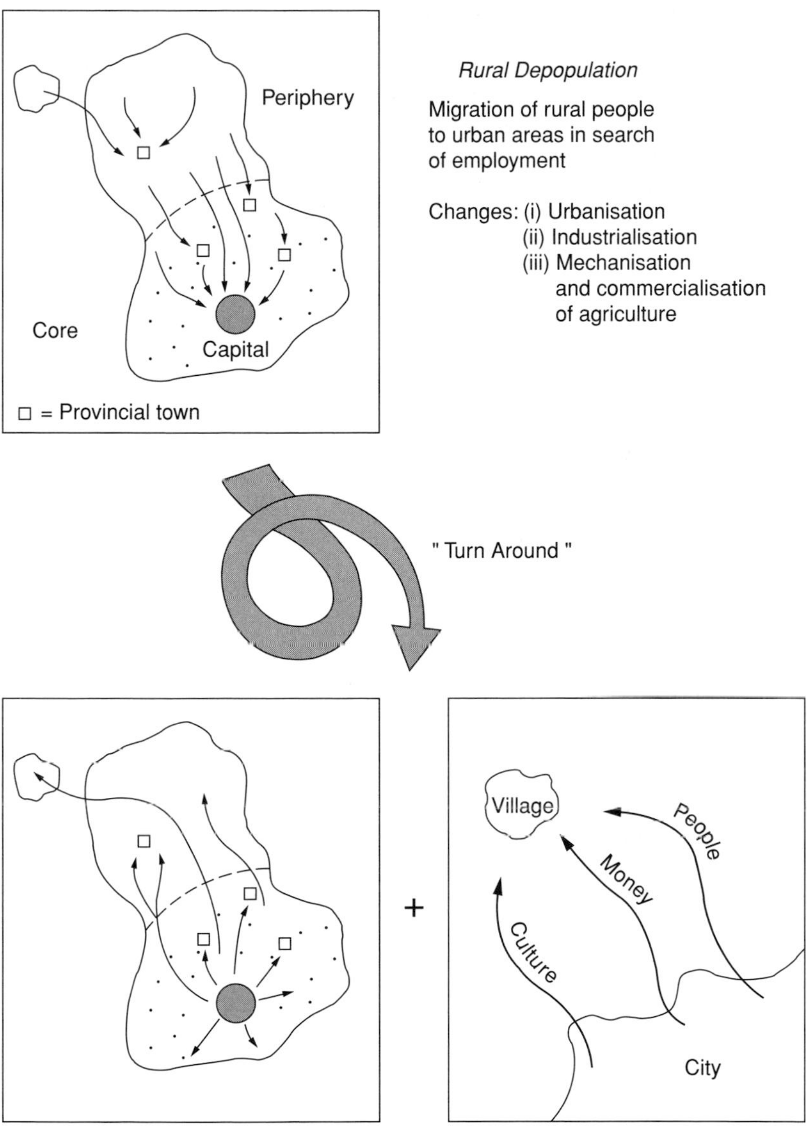

Figure 6.1 Demographic turnaround

much earlier, around the time of the 1901 census). The city failed to recover that figure in the post-war period because of the huge amount of its housing stock that had been destroyed in the blitz. In 1951 the population was 8.2 million; migration out of London had commenced. In the next three intercensal decades the population continued to fall: by 2.5% between 1951 and 1961, by 6.8% between 1961 and 1971, by 9.9% between 1971 and 1981. This last period with the heaviest outward migration coincides with the time in which counterurbanisation began to receive attention by geographers.

It has only been in the period between the censuses of 1981 and 2001 that London actually started to regain population. This increase in population through **re-urbanisation** has been the result of new developments such as those in the Docklands and other riverside areas especially in inner London, which have managed to attract a greater number of people to London than is leaving it. In the period between 1981 and 1991 re-urbanisation was at a very modest 0.9%, whereas between 1991 and 2001 it accelerated to 7.9%.

Two main parallel causes lay behind the counterurbanisation in south-east England: changes taking place in the CBD and inner city areas of London, and the opportunities offered by the areas beyond Greater London's boundaries. In the CBD, and particularly the City of London, as property values soared commercial functions continued to take over from other types of land use forcing residents out. As the need for office space expanded some of the CBD functions spread into selected well-positioned inner city areas, e.g. Finsbury and Clerkenwell, both within the Borough of Islington, forcing their property values up. At the same time many of the less favoured inner city areas, e.g. Clapton in the Borough of Hackney, were on the downward spiral of decline and being left with increasingly marginalised elderly and ethnic minority populations living in environmentally degraded locations with high unemployment and crime rates, and poor public service provision. This also had the effect of forcing the more mobile population away from the more central areas of London towards the outer suburbs and beyond.

Meanwhile the areas beyond greater London provided much more attractive prospects for potential counterurbanisers. From the late 1940s onwards, the string of New Towns located outside of London's green belt offered new housing and new employment. The New Town 'revolution' is still having its impact in south-east England in the twenty-first century with the continued growth of cities such as Milton Keynes (the largest of the New Towns with an eventual population of 250 000) and Peterborough one of the designated 'Extended Towns'. At another level, retirement to the traditional seaside resorts has been another element in the counterurbanisation of London. But in the last few decades it has been the villages and small market towns of the south-east that have attracted an increasing number of people away from Greater London.

What has happened to Greater London has also happened to the other main metropolitan regions of Britain, and at more or less the same time scale. The West Midlands, Greater Manchester, Tyne and Wear, Merseyside, South Yorkshire, West Yorkshire, Glasgow, South Glamorgan and Bristol all suffered from the combination of the loss of housing stocks during World War II and the closure of their heavy industries since the 1950s. New Towns and expanding towns initially helped to stimulate the process of counterurbanisation, but since then retirement to rural and coastal areas and then the movement of people of working age away from the conurbations towards more attractive rural districts and market towns have caused the process to continue. Unlike Greater London, none of the other large conurbations mentioned above has yet started to successfully re-urbanise by having more in-migrants than out-migrants.

Although most other northern European countries experienced counterurbanisation over the same time scale as Britain, from the aftermath of World War II onwards, southern European countries were much later in experiencing turnaround. Table 6.1 shows the total populations of the top ten Italian cities over the last six censuses. During the period from 1951 to 1961 all of the cities had rapidly expanding populations through urbanisation. There was an economic boom in the cities of the north and central Italy resulting from revitalisation of the older industries and the creation of new ones. Rome and most of the cities of the south had booming economies based upon the tertiary sector rather than manufacturing. Rural areas, especially in the south, were at this time not sufficiently productive to support their relatively dense populations. Massive rural–urban population flows were therefore sustained until the mid to late 1960s. The majority of the top ten Italian cities reached turnaround somewhere in the 1970s, with their populations peaking in the 1971 census. The economic boom gave way to a recession from

Table 6.1 Changes in populations of major Italian cities

	1951	1961	1971	1981	1991	2001
Rome	1.6	2.1	2.7	2.8	2.7	2.5
Milan	1.2	1.6	1.7	1.6	1.3	1.2
Naples	1.0	1.2	1.2	1.2	1.04	0.99
Turin	0.7	1.0	1.1	1.1	0.91	0.85
Genda	0.68	0.78	0.81	0.7	0.65	0.6
Palermo	0.49	0.58	0.64	0.69	0.68	0.65
Bologna	0.34	0.44	0.49	0.45	0.38	0.36
Florence	0.37	0.43	0.45	0.45	0.39	0.35
Catania	0.29	0.36	0.4	0.38	0.34	0.34
Bari	0.26	0.31	0.36	0.37	0.33	0.31

the mid-1970s and this slowed down the rates of rural–urban migration. Rome, because of its huge tertiary sector, continued to expand by attracting more in-migrants than people leaving it, and only reached turnaround during the intercensal period between 1981 and 1991. Likewise the two southern cities of Palermo and Bari, regional capitals surrounded by densely populated agricultural areas, continued to grow until the same intercensal period. Now all of these top ten cities are continuing to loose population to their surrounding rural areas. Italy's low birth rate and ageing population rates are also contributory to the process of counterurbanisation.

Within MEDCs, it can be seen that there is a clear sequence of which countries and cities have experienced counterubanisation at different stages in time: North America, followed by northern Europe and then by southern Europe. Counterurbanisation is still a rare phenomenon in LEDCs, but it is likely to affect them within the next few decades when rural–urban migration reaches saturation point and middle class families move away from city centres in sufficient numbers to achieve turnaround. Many LEDC cities, especially those in Latin American countries where urban crime rates are high, are developing outer suburban 'gated communities', which have often been described as 'ghettoes for the rich'; as these expand they will help to stimulate the momentum that will lead to counterurbanisation.

3 The Effects of Counterurbanisation

As counterurbanisation has influenced a growing number of regions within MEDCs, its effects have become more widespread. The effects of counterurbanisation are felt in rural and urban areas alike and can be summarised as follows.

a) The effects of counterurbanisation on rural areas

The overall effects of counterurbanisation on the countryside result from the influx of new people, the consequent population growth and the social, economic and environmental impacts brought about by these changes. Counterurbanisation puts a lot of pressure upon the rural environment, as well as blurring the edges between the rural and urban, both socially and economically. Some of the biggest changes that are taking place are:

- greater demands are being put upon farmland (greenfield sites) for housing and other developments
- small-scale manufacturing and service industries are being devolved into rural areas
- an increasing demand upon the countryside as a place for recreation
- the suburbanisation of villages (dealt with in detail later on in this chapter).

b) The effects of counterurbanisation on urban areas

Although not strictly within the confines of this book, the impact of counterurbanisation upon cities is important in understanding why people still leave them for the countryside. Counterurbanisation has an impact on all parts of cities through the ripple effect, but the areas that suffer from the most negative impacts are the inner cities as they have the greatest population loss. Some of the greatest effects are:

- the vicious cycle of decline or the reverse multiplier effect, which leads to the reduction of industries, shops and public services in particular areas as there is insufficient population or wealth to support them
- the marginalisation of the population as the more socially mobile move out and the poorer social elements such as ethnic minorities, single mothers and the elderly are left behind
- higher levels of unemployment as the number of industries and services decline
- the poor social and environmental fabric of these areas leave them as breeding grounds for a culture of drugs, crime and vandalism.

4 Industrialisation in Rural Areas

One of the most striking consequences of counterurbanisation has been the diffusion of industry into rural areas. Several factors lie behind these changes, which to some extent mark the return to the situation prior to the Industrial Revolution before industry in countries such as Britain became so concentrated in the coalfield areas and the large manufacturing cities. These factors are:

- the de-industrialisation of the traditional areas associated with coal mining and heavy manufacturing industries such as iron and steel and shipbuilding,
- the increasing amount of industry that has become **footloose** in terms of its possible place of location
- the lower cost of land for incoming industries
- the growth of high-tech industries with their favoured location on greenfield sites
- the very rapid growth of the small-scale enterprise employing fewer than 50 people
- the revival of some of the traditional craft industries in the countryside, particularly those that have their markets in the tourist areas (e.g. pottery, wood carving)
- the blurring of the boundaries between what constitutes secondary and tertiary industries has made it difficult to separate the two; many new service industries located in rural areas consequently bring some secondary activity with them

- the continued fall in the number of people employed in agriculture
- the movement of population from urban areas to rural areas increasing the size and range of skills of the rural labour force
- the availability of cheaper labour than in many urban areas (often casual, part time and seasonal).

Within Britain during the 1990s the process of diffusion of new industries had reached its final conclusion with the most rapid industrial growth taking place in the remoter rural areas rather than those most accessible to the large metropolitan zones. In the decade from 1990 to 2000, the county with the greatest increase in the percentage of industrial employment was Cornwall. Counties such as Cumbria, Norfolk, Herefordshire and Lincolnshire have also benefited from rapid small-scale industrial growth; all of these regions had the advantage of being designated Rural Development Areas in the mid-1980s and firms were given tax breaks and other incentives to locate within them. One aspect of the diffusion of industry into rural areas is the growth of the **rural business park** in the last few decades. These are the successors of the larger urban industrial estates that have their origins in the late 1940s and early 1950s. The key to their success has been their accessibility, being located not far from motorways or major trunk roads and their flexibility in having industrial units of a variety of different sizes.

CASE STUDY: THE CHEDDAR BUSINESS PARK, SOMERSET

Cheddar, in the Sedgemoor District of Somerset, was chosen in the District Local Plan as one of the growth centres for housing, services and industry. Although Cheddar currently has a population of 5500, making the size of a small town, it still rigidly adheres to its village status. Throughout the last few centuries it has had a certain amount of small-scale industry, mainly associated with the processing of agricultural produce. In the eighteenth century, in addition to a number of farms manufacturing cheese and other dairy products, there were 13 water mills on the upper reaches of the fast flowing River Yeo. In the first half of the twentieth century it had industries as diverse as cider mills, cement works using the limestone from the local quarries, shirt manufacture, and the making of baskets and strawberry punnets.

Today, the main types of industry in the village are those which typically represent the process of rural industrialisation: they are small scale, very diverse, based on industrial estates and employ local people either in well-paid skilled jobs or as fairly low-paid manual workers. The first industrial estate came to the village following the closure of the Cheddar Valley Railway line in 1969. The railway track was dismantled and the old station developed into what is now called the Valley Line Industrial Estate, in the early 1970s. This estate now has units dealing in such diverse areas as information technology, a technical training school, motor tyre and battery distribution, and wholesaling of outdoor pursuit clothing.

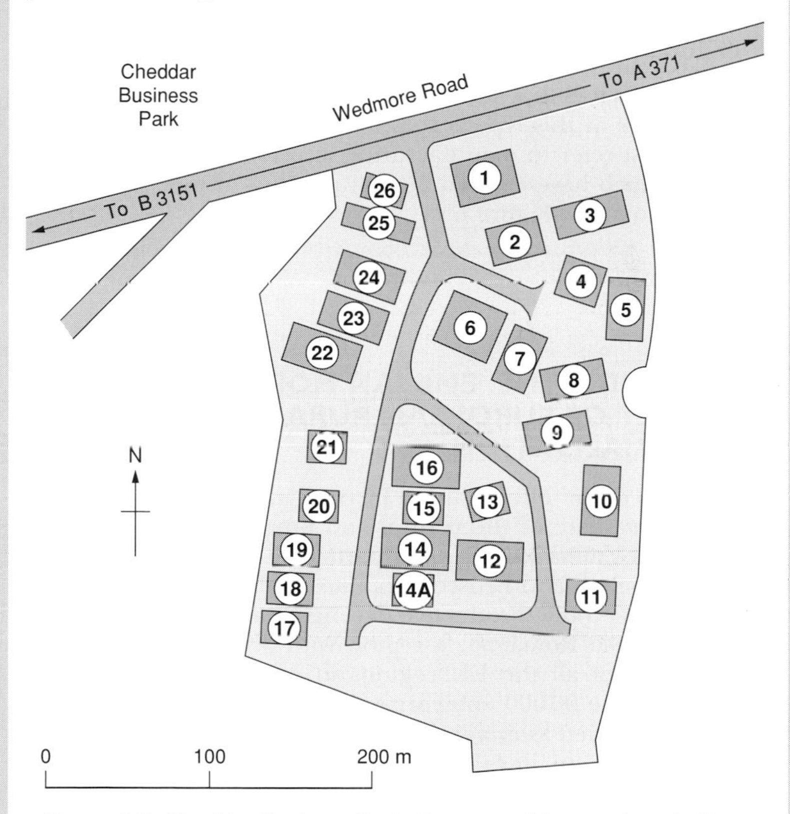

Figure 6.2 Cheddar Business Park, Somerset (the numbers indicate the individual units)

In the late 1990s a new industrial estate was set up in a flat site almost adjacent to the Valley Line Estate. The new development, known as the Cheddar Business Park, currently has 26 individual units and covers an area of 4.5 hectares. Provision is made on the District Local Plan for this estate to be almost tripled in size in the future. The 26 units have created an average of ten new jobs each, and by mid-2002 all but five were occupied. The range of businesses include some manufacturing, but most involve storage, wholesaling and distribution. Some of the wide variety of activities carried out in the Cheddar Business Park are:

- civil engineering and road surfacing
- electrical motor servicing
- shop fitters
- catering suppliers for public houses
- computer software for sewing machines
- ornamental wrought iron working
- bathroom fixtures and fittings.

One of the factors that comes most strongly out of examining an business park of this type is that 'industry' in this rural context does not just refer to manufacturing, but to a broader range of activities which have created a great number of new jobs in the tertiary sector (see Figure 6.2).

CASE STUDY: THE 'EMILIAN MODEL' AS THE CONTEXT OF EUROPEAN RURAL INDUSTRIALISATION

The 'trickle down' effect caused by rural industrialisation is well illustrated in some of the central and north eastern regions of Italy, and in Emilia Romagna in particular. In 1982, the Italian economist Sebastiano Brusco wrote an academic paper entitled 'The Emilian Model: productive decentralisation and social integration'. Emilia Romagna, a region with 3.9 million people is ranked 11th of all the EU regions in terms of its prosperity. There are some 90 000 enterprises in the region, and 41.5% of them are classified as *artigianati,* i.e. self-employed artisans. Over 90% of these small-scale industries employ fewer than 50 workers. This has enabled the industrial base of Emilia Romagna to be well diffused throughout rural as well as urban areas. For example, in the Carpi area where the specialist activity is knitwear production, over 50% of the firms employ fewer than nine people; many of these firms are effectively cottage industries located in rural settlements. As well as being based upon small family firms, the industries of Emilia Romagna are also frequently run as cooperatives, as this has been traditionally one of the most left wing regions of Italy. Of the nation's 43 000 industrial cooperatives, 15 000 are based in Emilia Romagna. The small-scale nature of these enterprises has enabled them to adopt up to date information technology with very few problems, especially as the region has high standards of education. It is not surprising that by the late 1990s Emilia Romagna was ranked as the second most prosperous of Italy's 20 regions.

The regions of Italy that follow the Emilian Model have often been labelled 'the Third Italy', distinguishing their small-scale industrial base from that of the northern 'Industrial Triangle' formed by Milan, Turin and Genoa, and that of the southern *Mezzogiorno* regional development area, both of which have been based primarily upon heavy industry. Whereas the other two industrial areas have suffered from the effects of recession and global competition, the regions that are characterised by the Emilian Model have been more resilient to economic change. The Emilian Model is therefore held up as an example to be followed in other parts of Europe, and certainly in those parts of Britain, where rural industrialisation is taking place, there appears to be higher survival rates for small businesses than is experienced in larger cities.

5 The Edge City: The Ultimate Manifestation of Counterurbanisation?

A growing phenomenon in the United States is the **Edge City**. This is essentially a product of counterurbanisation and suburbanisation in tandem. Pretty (1998) labels these settlements as 'Places with no History' and they provide the absolute contrast to the traditional villages and small market towns to which counterurbanising migrants aspire in Britain and elsewhere in Europe.

Joel Garreau in his 1992 book entitled *Edge Cities* gives a detailed description of their basic character, or lack of it. He explains the three waves of movements that have given rise to the phenomenon of Edge Cities since the 1980s: first came the movement of people away from the city centres and inner suburbs towards the urban periphery, then came the shopping malls and other services that followed them, and finally came the offices, financial services, high-tech industries and other employers. All of these involve low-density forms of building that have eaten up vast tracts of green spaces on the urban fringe. The inhabitants of Edge Cities still commute by car, but travel relatively short distances over new fast highways avoiding the other parts of the metropolitan areas of which they are satellites. The new roads have destroyed yet more of the countryside and as public transport provision is minimal, because of the relatively low-density populations to support them, private motor cars are a major source of atmospheric pollution. At the heart of the Edge City, as Pretty says, 'the new monument – no longer a cathedral – is the atrium, a climate controlled environment for the shopping experience'.

The Edge City is a place in which it is very practical and convenient to live, yet it has neither the benefits of city centre dwelling or those of country living. Around Greater New York there are already some 20 Edge Cities, but it is Southern California where the concept has become most developed. Within a 100 km radius of Los Angeles are 26 Edge Cities, and the concentration of them within Orange County has been compared to a theme park.

What happens on one side of the Atlantic is frequently echoed on the other, and in Jenkins' book *Remaking the Landscape*, Marion Shoard has written a contributory chapter entitled 'Edgelands'. This deals with the urban–rural 'interface' in contemporary Britain. As she suggests:

> 1 Between urban and rural stands a landscape quite different from either. Often vast in area, though hardly noticed, it is characterised by rubbish tips and warehouses, superstores and derelict industrial plant, office parks and gypsy encampments, golf courses, allotments and fragmented,
> 5 frequently scruffy farmland.

Although this suburban–rural no man's land is a far cry from the US 'Edge City' there are signs that the North American phenomenon is creeping into Britain by stealth. The large area on the northern fringe of Bristol known only as the 'North Fringe' is gradually taking on the characteristics of an Edge City. With a capacity to employ over 60 000 people, and rivalling the CBD of Bristol, this area has a vast out of town shopping centre (Cribb's Causeway), together with the new Ministry of Defence offices employing over 6000 people, a science park and the British or European headquarters of several very important high-tech industries. Amongst all these amenities are the vast areas of new low-density housing estates; the whole 'North Fringe' area is well connected to the rest of Britain by being close to both the M4 and M5 motorways.

6 The 'Exurbs' Concept

Another term connected with counterurbanisation and the growth of commuting in the United States is that of the **exurbs**. This concept was explained by Daniels in 1999, as a rural area with certain urban social and cultural attitudes, in many ways comparable to a suburbanised village in Britain. There are four conditions outlined by Daniels as to what constituted the exurban area:

- it is located 10–50 miles from an urban centre with a population of over 500 000, or between 5 and 30 miles from a centre with 50 000 or more people
- the commuting times taken by people going to work must be at least 25 minutes
- the community has a great mix of long-term and new residents

- the agriculture and forestry sector of the economy is still active, but declining in relative importance.

Some of the smaller states of the highly urbanised north-east coast of the USA have more exurban territories than those that are truly rural. Even some of the bigger states have a large proportion of exurban land. A survey of the different territories of Ohio carried out in the year 2000, shows approximately 30% of the state taken up by exurban land. One of the main reasons for this is that the main metropolitan areas of Ohio are well distributed throughout the state and a lot of its rural territory is within a 50-mile radius of a large city (see Figure 6.3).

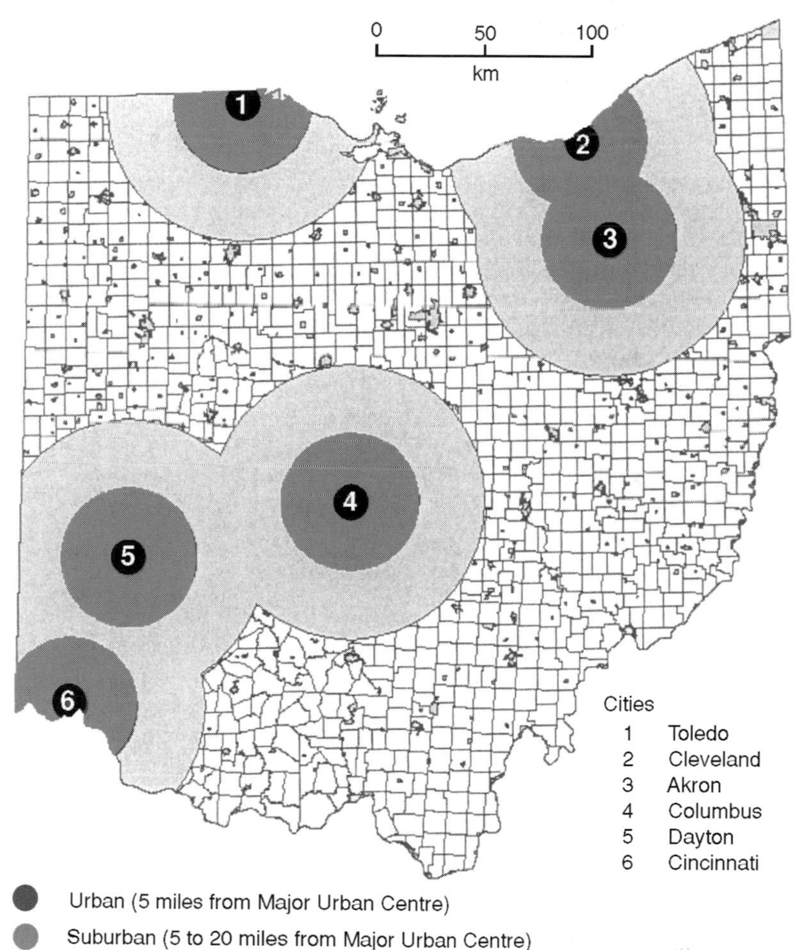

Figure 6.3 Exurban' areas of Ohio, USA

7 The Suburbanisation Process

Some geographical texts refer to the outward growth of a city's suburbs as **suburbanisation**. This definition not strictly correct. Just as urban growth is not synonymous with urbanisation, suburban growth is not the same as suburbanisation. Suburbanisation includes far more than the outward expansion of a city, as it also involves the impact cities have upon their commuter belts and the countryside beyond; this impact results from the processes of decentralisation and counterurbanisation.

Suburbanisation involves the changes that take place within rural settlements as commuter belts expand, more people buy second homes in the countryside and more people of pensionable age opt for a rural lifestyle. Within the UK, almost the whole of England is within the influence of the suburbanising effects of one or more cities; in Wales both the south with its proximity to Cardiff, Swansea, Newport and the urbanised areas of the old coalfield area, and the north with its proximity to Manchester and Liverpool are feeling the effects of suburbanisation. In Scotland, the central and eastern areas, are strongly influenced by similar trends from the major cities of Glasgow, Edinburgh, Dundee and Aberdeen.

Suburbanisation should not be regarded just in terms of the physical and economic changes that occur within rural settlements, but it also involves a complex series of demographic, social, cultural and perceptual changes. These changes now stem as much from the influences of the mass media, marketing and the globalisation of the economy as they do from the migration of people from urban areas to the countryside. In effect many rural dwellers in MEDCs are now *de facto* urban dwellers in terms of their culture and aspirations.

8 The Effects of Suburbanisation

There are many different but nevertheless closely linked effects that arise as the process of suburbanisation spreads its influence over ever-broader swathes of the rural environment. As mentioned above the process now affects most MEDCs and has started to have an impact upon some LEDCs. Figure 6.4 summarises these main trends taking place in rural areas in Britain, similar processes affect rural areas in other countries that are feeling the effects of suburbanisation. The main changes that villages undergo as suburbanisation takes place can be put into the five categories shown on the flow chart: demographic, economic, technological, social and cultural.

(a) *Demographic changes.* Decades of rural depopulation contributed to the ageing of the populations of rural settlements. The fittest and most able migrated away, leaving the older less mobile people behind. Following **turnaround** the demographic processes

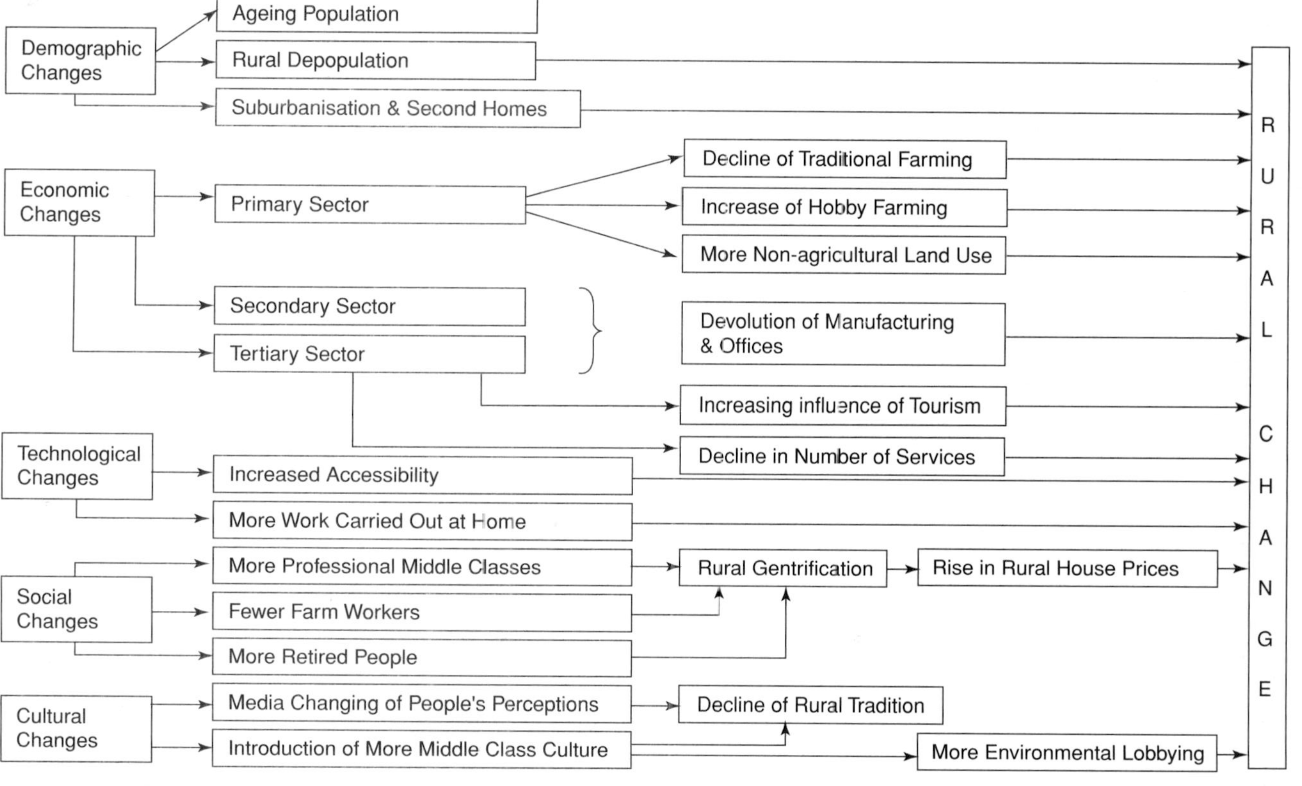

Figure 6.4 The effects of suburbanisation

did little to slow down the ageing of rural communities. The influx of retirees and second-home-owners (who tend to be in the higher age groups) have merely contributed to the proportion of older people in rural areas. Younger families moving into villages are unable to outweigh the older population as they are in much smaller numbers. Also villages must be put into the context that the nation as a whole is ageing. Only in large villages, or commuter villages close to expanding towns with younger populations, do rural settlements buck the national trend.

(b) *Economic changes.* These can be considered by sector.

i) The primary sector is undergoing the greatest changes. As farming has become more mechanised and more intensive fewer people have been engaged in agriculture (only 1.5% of the nation's workforce is currently engaged in farming). In recent years the reduction of subsidies, the changes in the EU Common Agricultural Policy and the onslaught of BSE and Foot and Mouth disease have further encouraged people to leave the land as a source of employment. A knock-on effect of these changes together with suburbanisation has been the conversion of farms to uses other than traditional agriculture these include: hobby farming, educational farms, rare breeds farms, and farms that combine agriculture and tourism.

ii) *The secondary sector.* This has rarely been of great importance in rural areas, but one of the effects of both counter-urbanisation and suburbanisation has been the devolution of smaller manufacturing industries into rural areas. This is discussed earlier on in this chapter.

iii) *The tertiary sector.* This is increasing in rural areas in two main ways. First, offices, wholesaling and other similar activities are more footloose than they were in the past, and they have relocated in rural areas for much the same reason as manufacturing industries. Secondly, the growth of the tourism and leisure industries in the countryside have also led to the expansion of job opportunities in the tertiary sector.

It is ironic that although new office-based activities may move into rural areas, villages are generally experiencing a loss of their traditional shops and services through suburbanisation. The more affluent newcomers to villages may not need to rely upon public transport, may send their children away to school or indeed may be childless and may have shopping habits related to their place of work. These situations may mean that there are insufficient villagers to support local bus services, village schools and village shops, and so the local service provision goes into decline. The elderly and the poor are those who suffer most from these changes.

(c) *Technological changes.* These are increasingly making the location of different occupations and activities more flexible. In theory improvements in transport make people more mobile and increase commuting distances. In practice, although road-building activity has been considerable in recent decades, this has been offset by increased congestion throughout the country. Rail improvements have been virtually non-existent and commuting times by rail appear to be getting longer rather than shorter.

It is in the realms of IT that the greatest technological impacts are being made. Computer technology enables firms to be increasingly footloose, and may attract them to locate in rural areas. At the same time an increasing number of people are now able to work from home by what is called **teleworking**, and they are able to opt for the less stressful rural environment.

(d) *Social changes.* These are largely brought about by the migration of newcomers into rural areas and the decline of the traditional farm labourer class. Suburbanisation brings more professional middle class people to villages, as well as the affluent second-home-owners and the retirees who are generally also relatively well off. This brings about the '**middle-classisation**' of rural areas and the **gentrification** of villages and other rural settlements. The wealthier new population invests in the improvement of their houses, and this leads to rising property values. This may be good for the investor but presents problems for young local couples who are first-time buyers on relatively low rural wages.

(e) *Cultural changes.* The cultural effects of suburbanisation are also associated with the higher economic status of newcomers and their social aspirations. The media have contrived to the breaking down of the differences between town and country by creating a national popular culture and by giving people the same material aspirations. These have helped to cause a decline in rural traditions. The influx of newcomers from urban areas has to some extent reversed this trend and stimulated a revival of crafts and traditions. Also the different perceptions of the countryside brought by the newcomers have led to more a active participation of people in environmentalist issues. The new populations may be less tolerant of proposed developments in rural settlements than the long-term population; there can be strong clashes of interest between these two distinctive groups.

CASE STUDY: THE IMPACT OF ROME ON ITS CAMPAGNA

Figure 6.5 is a conceptualised model of the changes that are currently are taking place in the countryside surrounding Rome as a result of both counterurbanisation and suburbanisation. The *Campagna*, the area of countryside surrounding Rome, was largely neglected and sparsely populated for about 1500 years following

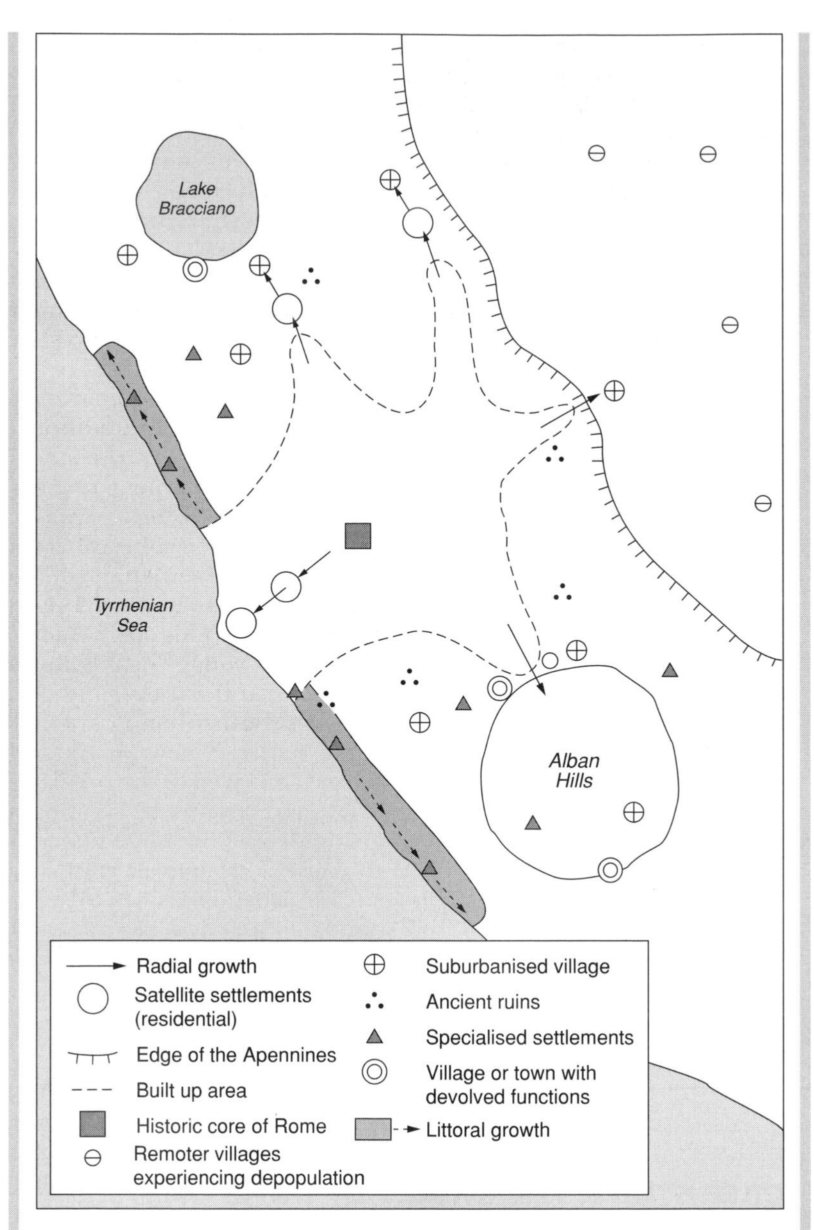

Figure 6.5 The impact of Rome's growth upon its Campagna

the fall of the city in 410 AD. There were vast tracts of malarial marshlands close to the coast, dense woods and scrublands on the hills and in the valleys of the interior, many abandoned Etruscan, Roman and later settlements lying in ruins, and general wildness and lawlessness that the citizens of Rome feared. The first accurate topographical map of the campagna was produced in 1834 by the British cartographer, William Gell and this large areas marked '*Solum imum Palustre*' (only marshland) and '*Regio non satis Explorata*' (region not well enough explored). This all reflects well the situation that continued until Rome started to expand from the late nineteenth century onwards when it became the capital of the newly unified Italy.

Continued urban growth has eaten into the surrounding countryside for the last 130 years, but this became particularly accelerated in the years of economic boom and heavy rural–urban migration when Rome almost doubled its population between 1951 and 1981. Much of this growth was distinctively radial and along the 'consular roads', i.e. the main roads constructed in ancient Roman times that are still the main roads into the city centre. As poorer migrants from the southern parts of Italy moved into the rapidly growing modern housing developments, being built speculatively on the farmland fringes of the city, some of the rich were moving out to live in villas built within the campagna. In the intercensal period between 1981 and 1991 Rome's population reached its 'turnaround' point and since then it has been undergoing counterurbanisation.

Many different factors have contributed to the movement of people out of Rome into the surrounding countryside:

- more people wishing to live in less polluted, congested and quieter environments
- the availability of many attractive environments close to the city: lakes, coastline, hills, mountains
- great improvements in the regional road and rail networks encouraging more commuting
- the construction of many new luxury suburban-type communities within the countryside (including 'gated' communities) either attached to or separate from existing villages or small rural towns
- the desire of an increasing number of rich and middle class people for second homes to spend weekends, but not too far from central Rome
- the devolution of some of Rome's functions into small towns of the campagna (e.g. the wholesale markets and the Bank of Italy have both moved out of town) creating new jobs
- new high-tech industries and quality workshop and craft industries been have established in various places such as in the Colli Albani to the south and the Bracciano to the north because of their attractive environments
- the problems of traffic congestion, pollution and crime continue to provide 'push' factors encouraging people to move out of Rome

- more rigid planning rules and the establishment of a proper green belt have prevented the continued building of speculative housing on the much of the urban fringe (this green belt was dealt with in Chapter 5).

The model shows that a whole series of complex movements and changes are taking place in the interaction between Rome and its Campagna. Radial growth is still taking place along the main roads out of town. Within the city boundaries new suburban developments continue to be built along the lower Tiber valley towards the sea. Along the coast littoral developments continue as part of the suburbanisation process. Most of the traditional, historic villages within the commuter zone are becoming suburbanised either by the buying up and conversion of older properties or by accretion with new residential areas being attached to the old. Some entirely new residential areas are being built in the countryside. Some villages and small towns are rapidly expanding because of the new functions they are acquiring. Some of the individual rural settlements remain highly specialised in their function or become specialised (e.g. thermal resort, hunting reserve or winery). At the same time as all this expansion is taking place there is some rural depopulation taking place in some of the remote and less favoured villages, and there are still plenty of reminders of settlement abandonment in the past, many of which are now significant archaeological sites visited by large numbers of tourists.

CASE STUDY: THE EFFECTS OF SUBURBANISATION ON CARTMEL, CUMBRIA

Even though Cartmel is located on the edge of the Lake District National Park and is some 1½–2 hours from Manchester, its most accessible large conurbation, it shows many distinctive suburbanisation effects. The village has good communications with the rest of Britain, being close to both the London to Glasgow (West Coast) mainline railway and the M6, which has been partly responsible for the changes that are taking place. With a population of 1800, Carmel has the social and economic profile of a village that has a large proportion of holiday homes, retirees and commuters, all of which are typical of suburbanisation.

Demographically, the village is growing slowly, at a rate of about 5% per decade. Although the birth rate (11 per thousand) is lower than the death rate (39 per thousand), the population is growing through inward migration. The low birth rate is a reflection of the fact that Cartmel's population is ageing. A total of 37.1% of the population is over the age of 60, in great contrast to just 12.9% of the population in the under-15 age group. The low birth rate and ageing population are also reflected in the average household size, which is just 2.2.

The gentrified, middle-class nature of the village is also reflected in statistical information on Cartmel. In terms of housing tenure, 83% of the population either owns their own homes or are buying them on mortgages, and this is in contrast to 4.6% of households in local authority housing and 7.7% in privately rented accommodation. An analysis of census data carried out for Cumbria County Council in 1999 categorises households into six different categories from A to F. In the A category, which represents people 'established at the top of the social ladder – healthy, wealthy and confident consumers' – are 72.8% of all Cartmel's households. The only other element found in large numbers belongs to the D category, which represents 'workers in the middle of the social spectrum who own or are buying their own homes'. The numbers in the E and F categories, those with the most difficult lives, were insignificant. Current house prices also reflect these trends; large detached houses were in 2002 in the £300 000–£400 000 price bracket. This information underlines the degree of gentrification and suburbanisation that Cartmel has undergone in the last few decades.

One of the consequences of the demographic, social and economic changes that Cartmel has undergone is the nature and range of the service provision within the village. Below are the results of a functional survey of Cartmel carried out in 2002. The village's shops and services were found to be very different from what they had been in the past. In the central cluster of streets close to the Priory church and around the market square the range of services reflect the fact that Cartmel is now a tourist and residential village, rather than the working agricultural village it was some 50 years ago.

The range of functions found in the centre of Cartmel are as follows:

- four public houses/hotels (more than would be normal in a place of this size)
- three antique shops
- two bookshops (one antiquarian, the other second-hand)
- one restaurant and two tea rooms
- a heritage centre/museum in the Mediaeval gatehouse
- a cabinet maker
- three shops selling either crafts, gifts, country clothing or combinations of them

- the 'Village Shop', the only grocery shop is the centre, but it is a high-class delicatessen
- an architect's office
- the old 'Village Institute' that houses a hairdresser and an upholsterer
- the Post Office
- a newsagent–tobacconist–sweetshop

(outside of this central area there are just two more shops: a 'Spar' grocery shop and a shop selling soft furnishings, as well as primary and secondary schools).

From this list it can be seen that there are just a few shops and services that could be associated with a traditional working village: the pubs, the Post Office, the schools and the convenience stores. The rest are the types of activities supported by tourists, weekenders, second-home-owners and the suburbanised long-term residential population. For the last group shopping needs are not properly catered for and visits to neighbouring market towns such as Kendal are necessary for a large range of both household and luxury goods. For those who are not car owners this involves using the limited local bus services. Although there are four buses each way each weekday between Cartmel and Kendal, two of these are scheduled as school buses bringing secondary pupils in to Cartmel from surrounding villages. An extra bus runs on Wednesday, which is Kendal's market day. This highly limited service is typical of so many rural areas in Britain, where car ownership may be high, but those who are too elderly or too poor to own a vehicle are at a great disadvantage.

The Cartmel Village Society brought out several publications for the Millennium and one of them lists the shops and functions that were present in the village at various stages between 1900 and 1950; the numbers are based on different operators at different stages in the half-century and they were therefore not all present at the same time. Most of this list is in sharp contrast with the range of functions found today and includes: five confectioners, three greengrocers, eight newsagents/stationers, four drapers, six bakers, five butchers, three cobblers, nine grocers, three banks.

What is clear from the list is that Cartmel was much more self-sufficient in its functions and services, it was acting as a market centre not just for its own population, but also for the surrounding district and there was generally much more 'everyday' commercial activity of the sort associated with the type of rural lifestyle that England had 50 years ago.

9 The Post-Modern Perspective on the Countryside and the Concept of the 'Post-Rural'

The economic and structural changes that are taking place in agriculture, together with the processes related to counterurbanisation, are totally changing the nature of the countryside in MEDCs. This is most marked in countries such as Britain where so few people are still engaged in agriculture (1.5% of the working population). Since the 1980s Britain has shifted away from traditional manufacturing industries, and has consequently become regarded as a **post-industrial** or '**post-Fordist**' society (the latter expression being a reference to the mass production manufacturing system introduced by Henry Ford). There have been similar great changes since the 1980s in the food production sector that need to be put into a historical context.

Agriculture has gone through a series of distinctive phases in the last 250 years in Britain. Until the Agrarian Revolution (which took place around 1750–1850) the agricultural economy was based mainly on the concept of local or regional **self-sufficiency**, where each village produced its own food supply and any surplus could be sold on to the neighbouring market towns for profit. The changes brought about by the Agrarian Revolution were to lead to the **commercialisation** of agriculture, which saw food being produced more efficiently for sale in local, regional or even national markets. The achievements of the nineteenth century were built upon throughout most of the twentieth with the continued mechanisation of agriculture and the wider use of the application of scientific methods to food production. In the period between the late 1930s and the 1980s, responding to economic realities and to government intervention policies, agriculture went through a phase of **intensification**; this was spurred on initially by the need to become more nationally self-sufficient in food during World War II and then continued for rather different commercial reasons. In the quest for sustained profitability, farms became larger, field sizes increased in order to make more efficient use of machinery, there was a greater application of chemical fertilisers and pesticides, as well as enhanced scientific selection of plant strains and animal breeds. All these factors led to the significant increase of yields per hectare.

Since the mid-1980s there has been a retreat from intensification that has been brought about by a variety of factors, the most important of which has been the EU's **Common Agricultural Policy (CAP)**. Overproduction of a whole range of foods, including meat, grain, wine and dairy produce in the 1980s in Europe, led the EU to bring about reforms in order to take land out of production through what is known as **set-aside**. Although uneven throughout Britain, the proportion of land that has to be taken out of production is as high as 20% in some of the more fertile parts of southern England. Three of the main changes taking place are moves:

- from **intensification** to **extensification**; this has enabled farmers to turn land over to less intensive activities such as increasing the amount of grazing land and it is having profound consequences in restoring habitats and biodiversity as well as reducing the amount of pollution from chemical fertilisers and pesticides
- from **specialisation** to **diversification**; the poor profit margins that many farmers have been achieving in the last two decades have led to increasing diversification into areas other than traditional food production, e.g. tourism and recreation, rare breed farms, educational farms – all these have implications on change of land use and land use intensity
- from **concentration** to **dispersion**; this would involve the reduction of the scale of some farms and the breaking of them down into smaller units, thereby spreading food production into some new areas; however, the continued trends that encourage 'economies of scale' have mitigated against much change in this direction thus far.

These various changes have, like de-industrialisation, led to what has been labelled a **post-productivist** countryside, i.e. one in which food production is not necessarily the most important activity. As productivism is associated with the modern age, post-productivism is associated with the **post-modern** age. The post-modern countryside has been given much attention by writers such as Ilbury (1996) and Cloke and Little (1997). Ilbury recognises four types of countryside in contemporary Britain.

- The **preserved countryside**, which can be found both in the conservation areas such as National Parks, AONBs (Areas of Outstanding Natural Beauty), SSSIs (Sites of Special Scientific Interest) and other nature reserves, as well as in the traditional rolling farmland areas which have retained their farmsteads, hedges and trees, and have therefore changed little in the past 250 years.
- The **contested countryside**, which is where the full effects of counterurbanisation are being felt; commuter developments, new transport links and industrialisation are all contesting with agriculture and other traditional types of rural land use for space.
- The **paternalistic countryside**, which is land that still lies in the hands of large estates; these lands are undergoing some change of function and selling of assets, but are not under the same threats as the preserved or contested areas.
- the **clientelist countryside** is associated most with the remoter upland areas where farming activities are more marginal and farmers still have to rely on grants and subsidies to keep themselves economically afloat.

Some authors have gone as far as labelling the twenty-first century countryside as being '**post-rural**'. This concept includes the post-modern, post-productivist countryside with the decline of its traditional food-producing role, but takes it one stage further. The concept of the post-rural is well identified in the works of various Canadian geographers, including Park, Coppack and Hopkins. Hopkins (1998) examines the 'post-rural' countryside in the research he carried out in the Great Lakes region to the west of Toronto, Canada. He explains how in this region the 'genuine countryside' with its agricultural roots and historic past is becoming commodified as a place for rural tourism. Through advertising slogans, logos and the creation of myths about the area, it has become part of a marketed and symbolic countryside, packaged and sanitised as a commodity to be 'bought' by the urban consumer at the weekend or when on holiday. This is where the boundaries between the real rural world, theme parks and virtual reality all become blurred.

In his research, Hopkins analysed the logos and advertising slogans of 210 pieces of promotional literature used by various tourist enterprises along the shores of Lake Huron. He found, by categorising the literature into the four headings of: natural environment, agriculture, heritage and recreation, that the messages put over were very different from the reality of the original countryside. Nature is portrayed as domesticated and docile by frequently using Disney-style cartoons to represent wildlife; agriculture is represented by historic farm buildings rather than the highly mechanised modern Canadian farms; heritage is often represented by the safety of the traditional family home environment and recreation is much more frequently portrayed as traditional sitting around the camp fire rather than as strenuous adventure pursuits.

Increasingly this type of myth creation is taking over rural and heritage tourism in MEDCs, and particularly within the countryside deeply influenced by its proximity to large cities. As Park and Coppack (1994) have put it:

> The creation of an idealized [*sic*] and innocent image of past rural ways and structures that somehow exists within reach of major urban areas reflects broader processes that have been linked in the wider geographical literature to post-modernity, post-Fordism and post-industrialization [*sic*].

Questions

1. Define the terms 'counterurbanisation' and 'suburbanisation', and consider the conditions that are necessary for them to take place.
2. Why do rates of counterurbanisation vary so much through time and from country to country?
3. Examine the main effects that counterurbanisation and suburbanisation are having upon rural areas and communities in MEDCs.
4. What is meant by the term 'demographic turnaround'? Examine the changes that it brings about.
5. Explain why rural industrialisation and the development of 'Edge Cities' are both symptomatic of counterurbanisation in MEDCs.

Summary Diagram

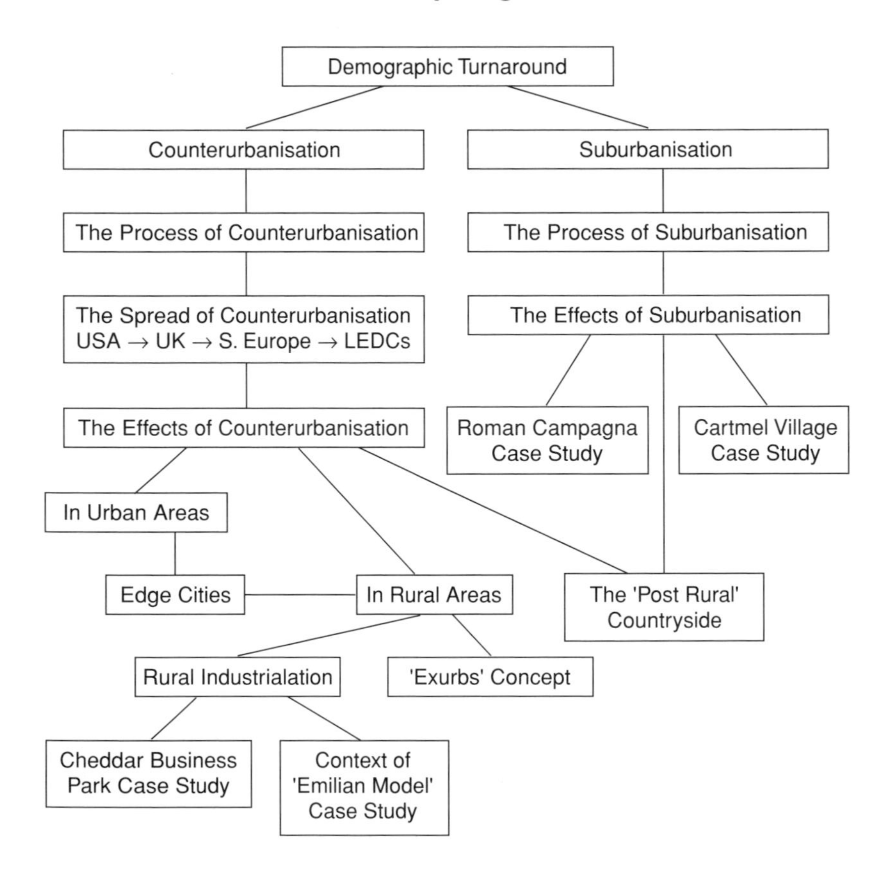

7 Changes and Conflicts in the Countryside

KEY WORDS

Village design statement: an initiative of the Countryside Agency to keep architectural and planning standards in villages in line with tradition.
New age travellers: a general term for people who have opted for a nomadic lifestyle in the tradition of gypsies.
DEFRA: the Department for Environment, Food and Rural Affairs (for England).
NDPBs: Non-Departmental Bodies: agencies of the government sponsored by ministries.
NGOs: Non Governmental Organisations: a general term for a wide range of organisations including charities and pressure groups.
Pluriactivity: carrying on more than one economic activity, e.g. farms that have diversified into tourism.
Oligopoly: a situation where a group of companies dominate the market.
Rural deprivation: a general term for the lack of services and opportunities experienced by rural communities.

> To every thing there is a season, and a time to every purpose under heaven: a time to be born, and a time to die; a time to plant and a time to pluck up that which is planted; ... a time to break down and a time to build up; ... a time to keep and a time to cast away
>
> *Ecclesiastes 3*

1 Change in the Countryside

Change is an ever-present part of rural life. Within the time span of the year there are the changes bound up with the passing of the seasons and it has been traditional practice for different agricultural activities to be allotted to each month or season. Within Europe the historical evidence is found in illuminated manuscripts such as the *Books of Hours*, as well as stained-glass windows and ceramics produced from the Mediaeval period onwards illustrating similar themes. These illustrations typically relate the farming year by showing the dominant activity on the land, month by month (see Figure 7.1). Today, although some of the more intensive agricultural practices in MEDCs, such as battery chicken farms, greenhouse cultivation of vegetables, flowers and fruit, and hydroponics, create artificial environments divorced from the outside world, most farming still has a close correlation with the changing seasons.

Figure 7.1 Ploughing, harrowing and sowing – picture depicting 'September' from a sixteenth century Flemish book illumination

In LEDCs, where the inputs from modern technology are more limited, the time of year remains closely associated with particular agricultural work. Even in equatorial locations, where there are few climatic variations in the course of the year, many crops such as the mango and cacao fruit seasonally rather than perennially and this has an impact upon the farming year and which activities are allotted to which months.

The longer term changes within the rural landscape are equally part of the normal progress of human development and adaptation to the natural environment, through a combination of social, economic and technological processes. Many of these changes have been dealt with in the previous two chapters and are connected with rural depopulation, counterurbanistion and suburbanisation as well as the shifts in the productivity of the land. In the contemporary world change is often taking place at a more rapid rate than in the past and this has great ramifications in the countryside, where society is generally more conservative than in cities. It is at the interface between rural and urban populations that conflicts of interest and confusion and contradictions in the perception of the countryside are most likely to occur.

2 Changing Perceptions of the Countryside

The collective perception of the countryside has varied dramatically from age to age as well as between places. In Britain and to a lesser extent elsewhere in Europe, one of the greatest contrasts in the perception of rural and urban environments is between that of the pre-industrial period and that of the period since the Industrial Revolution. For centuries in Britain, the clearing of forests and the extension of land under cultivation saw the gradual taming of the wildscape. Large tracts of countryside remained relatively wild right up until the Agrarian Revolution and the Acts of Enclosure, which took place from around 1750 onwards. In the pre-industrial age cities represented security, culture and learning; they contained all the positive features of a civilised society. By contrast, the countryside represented a wilder, less civilised environment with its dark forests, wild animals, diseases and outlaws.

It was only with the rapid expansion of the urban centres through industrialisation that these perceptions were turned upside down. In Britain, the huge, sprawling industrial cities of the North and the Midlands created very unhealthy environments for their inhabitants. Sub-standard housing, poor service provision, overcrowding and pollution from heavy manufacturing industries made city life barely tolerable for a large proportion of the population. Partly in reaction to the squalor of city life, attitudes towards the countryside and the natural environment were changing rapidly. Landscape painters such as John Constable and Richard Wilson, and poets such as William Blake, Samuel Coleridge and William Wordsworth, produced a new and romantic image of the countryside and rural life; this was the foundation of how the British still perceive their countryside today. In the collective mind of the British the rural environment is now associated with health, exercise and leisure pursuits, as well as being a place to just to go to in order to get away from the crowds and commune with nature. Of course, it is this perception of rural areas that has been one of the major stimuli for the process of counter-urbanisation; yet not all of urban–rural migrants find their idyll, so there are some counterflows back to the cities.

The ancient Romans had the concept of *urbs in rure*: the creation of a new town set within the countryside. In the nineteenth century, one of the side-effects of the problems of overcrowded industrial cities in Britain was the evolution of modern town planning and the concept of *rus in urbe* or countryside within the city. Philanthropists and enlightened industrialists created new industrial suburbs that took on the appearance of the traditional English village, with plenty of green recreational spaces, low-density housing and the services needed to create the spirit of community. One of the most successful of these is Bourneville, the suburb of Birmingham established by the Cadbury family for the workers in their chocolate factory between the 1880s and the early twentieth century. The settlement's centre is like the archetypal image of the idyllic English village. There is a village green with a covered market cross, a row of village shops, churches, schools, extensive playing fields and a community hall; even an old half-timbered manor house has been incorporated into the fabric of Bourneville. All that is missing is the village pub, a reminder that the founding fathers were Quakers.

3 The Contemporary Image of the Countryside: England and the USA

The perception of the English countryside that came out of the Romantic movement in the late eighteenth and early nineteenth centuries was merely strengthened throughout the twentieth century. So many poems, novels, pieces of music, films and television programmes have managed to create in the minds of the British an image of a beautiful countryside of rolling hills and carefully tended farmland, interspersed with historic half-timbered or stone-built farms and cottages. Poems such as A.E. Houseman's *A Shropshire Lad*, the novels of H.E. Bates, tales of childhood such as Arthur Ransome's *Swallows and Amazons*, some of Elgar's string music and Vaughan Williams' symphonies have all helped to build up the contemporary image of rural England. Films of Thomas Hardy's novels such as *Far from the Madding Crowd* and television series such as *Heartbeat* and *Midsomer Murders* as well as the TV dramatisation of H.E. Bates' *The Darling Buds of May* and the *Uncle Silas* stories have also contrived to give us an image of idyllic life in the country. This is the countryside of the chocolate box or the jigsaw puzzle. It is in fact a romanticised view of a rural England that never really existed. The thatched roof of the

Figure 7.2 *Helmingham Dell, Suffolk,* by John Constable – this romantic view of the English countryside and rural life still lingers today

idyllic cottage belies the fact that it provides a warm nest for rats and other vermin, the constant risk of leaking during winter storms and fire from sparking chimneys. The lives of the agricultural workers were also far from idyllic; it was hard toil from dawn to dusk with back-breaking labours, on low wages, few holidays and very limited career prospects.

This romanticised rural image has been used by many institutions and organisations. The National Tourist Boards of the component parts of the UK, the English Tourism Council in particular, and the regional tourist organisations have tended to use these well-worn images of the countryside to attract visitors from home and abroad. For a brief time in the late 1990s the images of trendy urban Britain were used to sell 'Cool Britannia' abroad, but soon proved unsuccessful and the more traditional images of heritage, rural landscapes and historic villages were readopted for advertising purposes.

Competitions such as 'Best Kept Village' and 'Britain in Bloom' have kept the concept of the pretty rural idyll going. In the last few years Millennium projects and National Lottery funding have enabled villages to add to their facilities and preserve their traditional environments. One such Millennium project has concerned village greens; it has enabled some villages to improve their central greens and to enable others to create new greens. One of the many roles of the Countryside Agency is to concern itself with **village design statements**. These involve careful planning, development and conservation of villages in order to enhance their atmosphere and to ensure that the materials used in new buildings together with other aspects of the village fabric are in keeping with the local traditions.

The collectively perceived image of the British countryside with its historic 'chocolate box' villages continues to inspire urban–rural migration; some surveys have shown that over half of town dwellers in Britain would aspire to move to the countryside if they had an opportunity. Now so much of the inputs into our perception come from the commercial world and the 'commodification' of the 'post-rural' countryside, as dealt with in the previous chapter. With parts of rural Britain becoming more and more like heritage theme parks, the idyllic myth has started to become a virtual reality.

Many social geographers have written about the contrasting attitudes and perceptions of the countryside between Britain and the United States. The USA, as a much bigger country with wilder landscapes and a much more recent history of agricultural colonisation by Europeans, evokes very different perceptions and emotions towards the rural from those in Britain. As in other countries, these attitudes are greatly in evidence in literature (e.g. J.F. Cooper's *The Last of the Mohicans*), painting (e.g. the works of the Hudson River School), photography (especially in the work of Ansell Adams) and, perhaps most of all, in the cinema. David Bell has written in Cloke and Little's (1997) book *Contested Countryside Cultures,* a chapter entitled 'Anti-Idyll: Rural Horror'. This deals with the portrayal of rural USA in a number of horror films. Bell contrasts the 'armchair countryside', which people may enjoy on their TV screens in Britain, with what he calls the *behind the sofa countryside* of the American horror film. In many films produced in the USA, the countryside is shown as menacing and its inhabitants threatening. Often it is the innocent city dweller, lost or stranded in the countryside, who naively becomes the victim of a sinister environment and its sinister people as in John Boorman's *Deliverance* (1972), Tobe Hooper's *The Texas Chainsaw Massacre* (1974) and Meiert Avis's *Far From Home* (1989). Another frequent theme is that of young people caught up in strange events taking place in a rural setting, such as Sean Cunningham's *Friday the 13th* (1980), Rob Reiner's *Stand By Me* (1986) and *The Blair Witch Project* (1998) – which was filmed by its actors.

What runs through so many US movies that use these rural themes are the concepts of the wilderness still being untamed, of the difficulties that faced the pioneers who opened up the frontiers still being present and of the violence involved in pushing back those frontiers lingering in the psyche of the rural population. Bell's conclusions about the rural societies portrayed in these films are:

Figure 7.3 In US films, the countryside and its inhabitants are often portrayed as sinister and untamed

Here are societies which seem idyllic, but which are malignant, at a dead end, and viewed in the grip of their own death throws. That condition is the real horror of the rural.

4 Other Rurals and Other Realities

Behind the facade of the 'chocolate box' village and the image of the idyllic way of life in the British countryside lie some very different realities. Many recent academic works on the countryside have looked at the less glamourous side of rural life. One of the most searching of these is Cloke and Little's *Contested Countryside Cultures*, which examines some of the very different realities that exist in contemporary rural areas. Rural poverty is generally less transparent than poverty in the inner city areas, which gets much more attention in both academic books and in the media. Certain groups of people who do not fit in with the general image of the ageing, white, genteel, middle-class rural population are equally lacking consideration in most works on the countryside. Some of these groups will be considered here.

a) The rural poor

Even though they might live in attractive environments of which urban dwellers may be envious, the rural poor may have great problems in making ends meet, and, as it has often ironically been pointed out, 'you cannot eat the view'. Rural poverty becomes hidden by regional statistics. As many rural areas become gentrified by retirees, commuters and second-home-owners who move in, those living close to the poverty line become concealed by figures that show rural districts and shire counties to be more prosperous than large swathes of the metropolitan regions. The families who have lived in the same village for generations, especially if they have worked as farm labourers, tend to be more vulnerable to poverty than the newcomers who may bring considerable capital with them.

Some of the main reasons for this vulnerability include:

- unemployment rates are higher than average, especially as some work in agriculture and in tourism is seasonal
- rural wages are generally lower than the urban equivalents
- house prices are relatively high, making them unaffordable to people with low incomes
- rural areas are poorly provided with public transport and this limits the mobility and employability of those who do not own cars.

Various studies carried out in rural areas of England have shown very high levels of poverty in certain counties. One carried out by Bradley in 1986 showed the following counties to have between 44% and 69% of their households within or on the margins of poverty: North Yorkshire, Northumberland, Nottinghamshire, Shropshire, Devon, Wiltshire, Warwickshire, Cheshire, West Sussex and Suffolk.

The invisibility of this rural poverty will be more extreme in such places as Cheshire and West Sussex, which are comparatively prosperous counties where, as Cloke and Little (1997) put it:

> [the problem] is vitally interconnected with experiences of powerlessness and marginalisation drawn from living cheek by jowl with people of affluence and status.

Female employment in rural areas is another contributory factor to the poverty of certain households. Women are generally on lower pay, less mobile and more likely to be involved in part-time work than men in rural areas. Cloke and Little (1997) carried out research in certain North Somerset villages within the commuter belts of Bristol and Bath and found that 52% of the women surveyed were doing work for which they were not using their educational qualifications. This reflects both the lack of availability of well-paid professional jobs in rural areas and the fact that the women in the villages surveyed, for one reason or another, were not as mobile as the men.

In the United States rural poverty is more clearly recognised than in the UK. Specific areas with a strong regional identity have long been recognised as having a problem of poverty and have therefore received considerable government help. They still, however, are less focused upon than the areas of urban poverty; this can be put down partially to the 'frontier' mentality in the American psyche, which perceives rural life as involving necessary hardships. Two regions of the United States that continue to have high levels of rural poverty, for

Figure 7.4 A modern view of the country in this cartoon from *The Times*

historical reasons are the Deep South, with its large concentrations of African-Americans, and the Appalachian Mountains with their unique communities and customs.

b) Ethnic minorities

The non-white ethnic minorities in Britain are essentially town dwellers. There are historic reasons for this, as the largest groups of Asians and Afro-Caribbeans came to Britain in the decades immediately after World War II in order to fill the vacancies in industrial and tertiary employment in the major cities. As each stream of migrants came to Britain, they became more concentrated in the specific parts of the cities where they were already established as a result of family and community ties, and the work of immigrant associations. The clustering of ethnic groups reflects both the sense of belonging to a community and to the security that the 'safety in numbers' concept gives.

The diffusion of non-white minorities into rural areas is slowly taking place, but thus far only in small numbers. Retirement is one way through which this diffusion is taking place, as well as second and third generations who have been educated in Britain and have become as mobile in the labour market as their white counterparts. The diffusion of ethnic restaurants down through the settlement hierarchy is another way in which non-white groups have found their way to small market towns and large villages. The exact levels of non-white populations in rural areas is unknown, but one estimate is that the ethnic minority groups make up about 36 000 of the population of Cornwall, Devon, Somerset and Dorset. Although racial violence has generally been restricted to urban areas in Britain, rural areas are more conservative and potentially less accommodating towards ethnic minorities. In the late 1990s there were some unpleasant racist scenes of aggression towards the small local Asian community in the Somerset market town of Yeovil. Rural areas close to ports and airports where asylum seekers are being held are also places of potential friction with the local communities; the Isle of Thanet in Kent, near the port of Dover, and the SW Essex area close to Stansted Airport are two rural locations where asylum seekers are being processed. In order to take some of the pressures off the influx of LEDC migrants into Britain, the government has considered the possibility of temporary visas for casual seasonal agricultural labour. This is already well established in some other EU countries, such as Spain and Italy.

c) Children

The images of an idyllic childhood in the English countryside are common in the literature of the first half of the twentieth century, with such books as Laurie Lee's *Cider with Rosie* and the Enid Blighton's *Famous Five* series. The idea of the countryside being a healthy environment for children to grow up in is indeed one of the most positive factors behind the counterurbanisation process. Although still generally better for children than large urban areas, growing up in the countryside today is unfortunately not as attractive as it was 40 or 50 years ago, and there are a whole variety of factors that have brought about this change:

- ageing of the population has reduced the number of children living in rural areas, especially in smaller villages and hamlets, and this has led to a degree of isolation and lack of companionship
- mechanisation and intensification of agriculture and stricter enforcement of the laws of trespass have left less of the countryside open for children to play and have adventures
- problems of predatory strangers and consequent parental fears of their children's vulnerability
- closure of small village schools, making children travel further for their education
- 'cultural urbanisation' of country living, giving children the same aspirations as in cities and therefore causing them to spend more time indoors with their computers, recorded music and friends
- pressures of advertising and peer pressure both encourage children to grow up more quickly and thereby lose an active interest in the countryside at an earlier age
- spread of what had been regarded as 'urban problems', such as drugs and crime, into rural areas has made them less safe for children.

d) Travellers

Romany gypsies have lived in Britain for hundreds of years and their nomadic lifestyle and brightly painted caravans became part of the romantic vision of the traditional rural way of life. Although their numbers have now dwindled into insignificance, since the 1960s and 1970s, various groups of people have taken to the road and emulated the Romany lifestyle. These people, who have dropped out of the 'normal' pattern of a sedentary society that works from nine to five, have done so for a whole variety of reasons. Generally given the name New Age Travellers (NATs), these nomadic groups may be motivated by a rejection of materialism, a love of nature, a dislike of the urban 'rat-race', the embracing of hippy culture or a combination of any of these attitudes.

One of the biggest problems that NATs face is their contact with country people and in particular with local landowners and councils. Although they often have attitudes towards the natural environment that may be very similar to those of villagers, they are frequently perceived as a menace, generating lots of rubbish, taking drugs and indulging in petty crime. The Nimby ('not in my back yard') attitudes of the rural population towards NATs has often been made worse by the lack of local provision of official camping sites for the travellers, making them resort to unofficial sites and then being moved on through police intervention. In 1995, The Criminal Justice Act made it easier for NATs to be prosecuted through a new offence of 'aggravated trespass'.

e) The lesbian and gay population

It is much easier for lesbians and gay men to be themselves within the anonymity of large cities. In rural areas with tight-knit communities, they are frequently forced to lead sheltered and dual lives. Given the small numbers of like-minded people in rural areas, as well as the lack of meeting places, and the social conservatism that can breed homophobia, young lesbians and gay men generally find themselves part of the rural–urban migratory flow. In the large cities they can be in touch with much larger communities and have at hand specialist pubs, clubs, listings magazines and help-lines.

In Bell and Valentine's *Mapping Desire* (1995), which was the first academic geography text to be devoted to human sexuality, Jerry Lee Kramer has written a chapter entitled 'Bachelor Farmers and Spinsters'. This deals with the lot of lesbian and gay farmers in rural North Dakota. North Dakota is one of the most sparsely populated states of the USA and this presents great problems for lesbians and gay men who wish to find partners. Kramer describes the courting rituals of the 'bachelor' and 'spinster' farmers, which involves driving long distances in their cars or pick-up trucks to specific known petrol stations where they can encounter other lesbian and gay people.

5 Conflicts in the Countryside

a) Land-use conflicts

Rural areas are rarely without conflicts of interest. As countries become more urbanised, there are greater pressures imposed upon rural areas for an increasing range of activities. As has been seen in previous chapters, urban sprawl and the changes associated with it are imposing pressures upon the land. Within rural areas, there continue to be conflicts between the various traditional forms of rural land use. As well as land being used for food production, there is pressure from the need for rural housing, new roads and other forms of infrastructure, mining and quarrying, forestry, water supply and conservation, and the leisure and tourist industries. The conflicts are not just about space but also involve the issue of employment, as the various non-agricultural activities may be able to bring badly needed jobs to rural areas.

Rural land-use conflicts in Britain are probably most strongly felt in the National Parks. Here the competition for space is very great and has to be brought into check by the needs for conservation. Although the biggest areas of potential conflict in National Parks are probably still the same as when they were first established – access by the general public to land and the rights of way issue – there appear to be an ever-increasing number of conflicts.

Easy access to the parks by car has become a major issue. The Peak District National Park has a particular problem in being wedged between Manchester and Sheffield, and the sheer volume of traffic at weekends has led to the closing of certain popular scenic roads such as that around the Ladybower, Derwent and Howden Reservoirs. In other National Parks 'honeypot' sites have also become subject to restricted car access, e.g. the area around Haytor on Dartmoor.

Many of the National Parks, as they are located mainly in the Highland Zone of Britain, have valuable stones and minerals within them. In some cases quarrying and mining still take place within the parks, in others the park boundaries are manipulated so that these geological eyesores are excluded, e.g. Blaenau Ffestiniog slate quarries in Snowdonia. Forestry and reservoirs can enhance landscapes, but conservationists frequently criticise them as they are not natural but man-made environments, e.g. the Keilder Water area of the Northumberland National Park. Until recently most planted forests in National Parks were in the form of regimented rows of non-native conifers, inside which – if public access is not limited by the Forestry Commission – visitors would find very limited biodiversity in comparison with native deciduous woodlands. The flooding of vallcys to create reservoirs also can have the effect of reducing the local biodiversity.

The wildness of the National Parks has made them valuable places for army training; this too can create another form of land-use conflict. Visitors to Dartmoor, and the Northumberland and the Brecon Beacons National Parks find themselves excluded from 'danger areas' when live ammunition is being used for firing practice.

b) Conflicting-interest groups

A very large number of organisations are involved with what goes on in the countryside. Some of these have functions that overlap, others may be in direct conflict with one another. In Britain the organisations can be put into three categories: governmental institutions, NDPBs (Non-Departmental Public Bodies) and NGOs (Non-Governmental Organisations), which include pressure groups and associations supported by membership.

The governmental institutions operate at both the national and local levels. Each country of the UK has a separate ministry in charge of rural affairs: DEFRA (Department of Environment, Food and Rural Affairs) in England, SEEFRAD (Scottish Executive Environment and Rural Affairs Department), ARAD (Agriculture and Rural Affairs Department) in Wales and DARNI (Department of Agriculture and Rural Development, Northern Ireland). Other government departments are involved in rural areas and their development, e.g. the Regions and Transport.

The Forestry Commission counts as a separate government department and is responsible for the protection and expansion of woodlands. The Food Standards Agency, whose work was once part of the old Agriculture ministry, now also counts as a separate government department. At the local government level there may be up to three layers of administration. In many shire counties there are three tiers formed of parish councils, district councils and county councils; when these are of different political persuasions there is plenty of potential conflict.

NDPBs are closely linked to ministries but are strictly not part of them. A whole range of important bodies are sponsored by DEFRA and these include:

- the Countryside Agency, which deals with a wide range of issues pertaining to the countryside, rural communities, conservation and development
- the Environment Agency, which deals with rivers, water quality, flooding and flood defences
- English Nature, which concerns itself with the conservation of wildlife and natural landscapes. As with many other agencies, it has parallel organisations in Scotland, Wales and Northern Ireland.

There are numerous NGOs and pressure groups involved with rural areas and rural issues. Some of these organisations are actively supporting farmers and rural communities, and helping them to overcome the rapid changes that may be taking place. The National Farmers' Union, the Royal Agricultural Society, ACRE (Action with Communities in Rural England) and NREC (the National Rural Enterprise Council) all work for an active and living countryside. The NREC is committed to the spread and application of information technology in rural areas and has set up an important on-line information service, 'InfoRurale'. Support for farmers and others who are in financial and other difficulties is provided through the charity 'The Rural Stress Information Network'.

NGOs concerned with conservation and rural pursuits may find themselves in conflict with farmers and other landowners, but normally they should be able to work together. Conservation organisations include the National Trust which not only buys and restores country houses and their estates as well as other historic places, but also is one of the biggest landowners of areas of outstanding scenic value, including hundreds of kilometres of coastline. A much smaller organisation, the Countryside Restoration Trust, is involved in the preservation of old farm buildings and traditional farming areas under threat, such as water meadows. The CPRE (Council for the Preservation of Rural England) is concerned that no large-scale developments should take place in rural areas and runs various campaigns such as those against building housing on green field sites and the further intrusion of airport noise into the countryside.

The Ramblers' Association (RA) is one of the most powerful pressure groups dealing with outdoor activities. They campaign for greater public access to the countryside and in doing so may come into conflict with landowners. The RA was in fact the most influential group in the movement for national parks as a result of their 'mass trespass' of Kinder Scout in the Peak District in the late 1940s. The Countryside Alliance is involved in many aspects of rural life, but one of its main campaign areas has been in support of the various forms of hunting. Organisations such as the League Against Cruel Sports and other pro-animal lobbyists are diametrically opposed to the aims of the Countryside Alliance.

6 Is There Really a Crisis in the British Countryside?

A lot of attention has been given in recent years to the problems that are being experienced in rural areas in Britain as a result of economic change. Some of these issues connected with counterurbanisation, suburbanisation and the evolution of a 'post-productivist' countryside have already been discussed in the previous chapter. Two issues in particular need further attention: the changes that are taking place in agriculture and causing difficulties for the farmer, and the changes that are being brought about in rural service provision.

a) Problems of change in agriculture

The declining number of people engaged in agriculture and the difficulties a large number of farmers are experiencing in making a living are regarded as one of the main issues of the perceived crisis in the British countryside. Some of the structural changes that are taking place in agriculture were dealt with in the last section of the previous chapter.

These changes, which can be put into three distinct historical phases, are summarised in Table 7.1. From the 1920s until the 1980s British agriculture followed the pathway towards intensification and higher yields per hectare. This was achieved in many ways, including: the putting of marginal land into production, mechanisation, the increased use of chemical fertilisers and pesticides, the merging of farms, the increasing of field sizes. The movement towards intensification was encouraged by government action through such channels as subsidies on various products, e.g. milk and grain. Subsidising continued under the European Union Common Agricultural Policy until the overproduction phase was reached in the 1980s. Since when there has been a retreat from intensive agriculture and towards land being put out of production and into 'set-aside', a more sustainable and 'environmentally friendly' approach to farming and farmers being encouraged down the road to **pluriactivity** (diversification into other sectors such as tourism)

Making the adjustments from intensive farming with subsidies towards more diverse forms of rural activity has not been easy, and various changes and crises have contrived to make the situation even more difficult for farmers. These changes have included:

- *The role of supermarkets in the food chain.* Six supermarket groups control about 52% of the food market as an **oligopoly**. This enables them to have a great leverage over what farmers get for their produce, especially as they can flood the market with cheaper foreign imports. Currently, farmers get on average around 7p in the pound for what is sold in the supermarkets, whereas in the early 1990s they got 15p in the pound.
- *The BSE crisis.* In the mid-1980s Bovine spongiform encephalopathy (BSE) was first identified in cattle in Britain. This condition that leads to the irreversible deterioration of brain tissue appears to have been spread by the recycling of the remains of cattle brains in cattle feed. The disease was then passed on through the food chain to humans as vCJD (new variant Creutsfeld Jacob Disease). Over 12 700 cattle in Britain caught BSE and 122 people have so far died of vCJD. Scientists do not know the incubation period of vCJD, if it is as much as 60 years some 135 000 people could be at risk, but it is hoped that it will be much lower. BSE did enormous damage to the home market sales of beef and export sales ceased for a long time to most other European countries. Cattle aged over 30 months (which are believed to be more susceptible to BSE) are still not used for beef.

Table 7.1 The main 'food regimes' in Britain

Character	First regime	Second regime	Third regime
Products	Grain, meat	Grain, meat durable food	Fresh organic, reconstituted
Period	1870 to WWI	1920s to 1980s	1990s
Capital	Extensive	Intensive	Flexible
Food systems	Exports from family farms in settler colonies	Transnational restructuring of agriculture to supply mass market	Global restructuring, with financial circuits linking production and consumption
Characteristics	Culmination of colonial organisation of precapitalist regimes Rise of nation states	Decolonisation Consumerism Growth of forward and backward linkages from agriculture	Globalisation of production and consumption Disintegration of national agrofood capital and state regulation Green consumers

- *Other food safety issues.* The BSE crisis left the British population more cautious about the safety of food, particularly meat. Outbreaks of swine fever and *E. coli* food poisoning have contributed to the general public's suspicion about animal welfare and the treatment of meat in butchers' shops. The conservative and cautious public opinion is also strongly against experimentation with GM (genetically modified) strains of crops.
- *The strength of the pound.* For the last ten years the pound sterling has been generally overvalued in relation to other hard currencies, especially the dollar, the Deutschmark and the French franc. When the euro was launched in January 2002, it fell in strength very rapidly against sterling. This made British goods expensive in the major overseas markets and left UK agricultural produce uncompetitive. Entry into the Eurozone could make the situation much better for British farmers.
- *The foot and mouth disease (FMD) crisis.* When British agriculture was just recovering from the impact of BSE, an epidemic of FMD broke out. The epidemic started in Northumberland but spread rapidly to other parts of the country through the movement of livestock to markets and slaughterhouses. Some of the worst affected counties were Cumbria, Devon, North Yorkshire and County Durham in England, and Dumfries and Galloway in Scotland. In the period between February and August 2001 over 3.3 million animals were slaughtered. Not only was this a source of great distress to the

farming community, but it also involved the cordoning off of huge tracts of land, including the closing of Dartmoor National Park to the public. The FMD outbreak gave very bad publicity about Britain to the outside world as huge funeral pyres of animal carcasses were televised worldwide. Not only did FMD have a huge impact upon farmers' incomes (thousands of farmers left the land altogether after the experience), but also did enormous damage to tourism in Britain.

- *The banning of various forms of hunting.* The debate over the various forms of hunting with dogs and other bloodsports has given even more impetus to the idea of the countryside being in crisis and of the misunderstanding of country life by city-based politicians. Although emotions run high over this issue, the amount of rural unemployment that the end of fox hunting and stag hunting will cause is fairly limited; however, individual communities such as those on the higher parts of Exmoor will be badly affected.

The Department for the Environment, Food and Rural Affairs (DEFRA) in England, and its counterparts in Wales, Scotland and Northern Ireland, all have various initiatives and schemes to encourage new agricultural developments, for which farmers are awarded grants and other incentives. These are intended to help farmers out of their financial difficulties, to enable them to cope with change and to create a greater harmony between farming and the natural environment than there has been in the recent past. DEFRA has a series of schemes collectively called the 'English Rural Development Programme' (ERDP). One of the main reasons for these initiatives has been, to quote DEFRA, to 'play a major role in helping rural communities recover from Foot and Mouth Disease'.

Some of these initiatives were established in 2001–2, but others had already been operative since the 1990s and were given greater prominence. They include the following.

- The Countryside Stewardship Scheme is the most important of these incentives that sets out to enhance and restore special landscapes, habitats and historical sites. An example of the success of this scheme has been the restoration of over 1300 km of dry stone walls and 8000 km of hedgerows.
- The Environmentally Sensitive Areas Scheme (ESAs) encourages farmers to adopt methods that will safeguard certain areas of great landscape, wildlife or historic value. So far 22 ESAs have been established and include such diverse landscapes as the Penwith Peninsula in Cornwall, Exmoor, the Brecklands of East Anglia, the Somerset Levels and the Cotswolds.
- The Organic Farming Scheme encourages farmers to convert their land to organic methods of crop production, which both protects the soil and meets the increasing demands for organic produce.

- The Energy Crops Scheme encourages farmers to coppice woodlands, new and old, in order to grow timber to be used as fuel for heating and for electricity generation.
- The Hill Farm Allowance Scheme helps farmers and the rural communities in marginal upland areas to continue their traditional forms of agricultural land use and way of life.
- The Vocational Training Scheme gives assistance to local communities with training in agriculture, forestry and other rural occupations in order to increase their skills base.

b) Problems of rural service provision

Chapter 6 looked at the processes of counterurbanisation and suburbanisation that have been responsible for the decline of service provision in British villages. The changing economic and social situations in rural areas have led to this decline. Smaller numbers of people working locally and particularly in agriculture, and more commuters and non-resident second-home-owners have meant that there are smaller populations to support village shops and services. At the same time **economies of scale** have made village shops uncompetitive in comparison with the supermarket chains in neighbouring market towns and **rationalisation** has led to the closure of some services such as post offices and primary schools in smaller communities.

The Countryside Agency carries out surveys of rural service provision. The most recent detailed survey in 1997 found that:

- 28% of parishes had no public house
- 42% of parishes had no permanent shop of any kind
- 43% of parishes had no post office
- 49% of parishes had no school
- 75% of parishes had no daily bus services
- 83% had no GP based in the parish
- 92% had no police station.

Statistics like these are indicators of **rural deprivation** and it can be argued that this takes various forms, two of the most important being **the deprivation of opportunity**, whereby people living in rural areas are increasingly limited in their career prospects, and **mobility deprivation**, which means that because of transport problems and isolation people do not have access to jobs elsewhere. Many of the charities and other interest groups mentioned earlier in the chapter are trying to address these problems associated with rural decline.

The Countryside Agency itself has numerous schemes and initiatives aimed at improving the lot of rural communities, these include:

- defining 'Rural Priority Areas' where action needs to be concentrated; these currently take up about 35% of the land area of England
- the Rural Transport Partnership Scheme aims at dealing with the problems of poor transport provision and finding solutions, which

may include minibuses, community taxis and dial-a-ride schemes in areas where large buses would not be sustainable

- the Village Shops Development Initiative aims at improving and modernising existing village shops and providing grants for people to set up shops in villages that are deprived of them
- the Land Management Initiatives encourage the partnership between farmers and the local communities to develop sustainable land management systems that contribute to an attractive environment
- the Local Heritage Initiatives enable local communities to research and take care of their landscapes, landmarks and traditions.

7 Rural Decline

The problems faced by agriculture and rural areas in the UK are a reflection of the changes that are taking place throughout Europe. In 1957 there were some 22 million farmers in the EU. Since the establishment of the Common Agricultural Policy and the structural adjustments that it brought about, the number of farmers has declined to just under 7 million. The total number of farms in the EU has declined even more rapidly: between 1989 and 1995, 1.3 million farms disappeared through land holding consolidation, urban sprawl or just by becoming uneconomic. The demographic changes in the rural areas of the EU are also causing problems as the average age of 50% of EU farmers is now over 55.

The ways in which these changes have influenced individual countries have varied considerably, as land tenure, holding size, efficiency of production, marketing and transportation all varied greatly from country to country. The European Union's policies towards improving the lot of the countryside dweller are both varied and complex. There is felt to be need for a form of positive discrimination towards rural areas as their average wealth is only 83% of the EU average.

There are two prevailing theories and approaches to rural development, the so-called **endogenous model**, which seeks to build upon and develop existing resources within a rural area, and the **exogenous model**, which relies upon outside investments and expertise to modernise the countryside and its economy. In Europe and within the practice of the European Rural Development Commission, it is the exogenous approach that is commonly used. Of current regional development funding by the ERDF (European Regional Development Fund) only about 10% of investments go to rural areas. Five types of priority areas have been defined by the EU regulations and of these only two include rural areas. The areas most supported by EU funding are the most marginalised rural areas of Europe (e.g. Ireland, Greece, the Italian *Mezzogiorno*, Corsica, and the Highlands and Islands of Scotland).

Rural life in the EU member countries has a long tradition and rural development policies are designed to safeguard the best of the traditional values, but at the same time to equip rural communities for survival in the modern world economy. Some of the most important aims of the EU Rural Development Commission are:

- to make the agricultural sector competitive in the global market
- to encourage specialisation and diversity in agricultural production
- to modernise holdings and systems of land tenure
- to encourage environment friendly systems of production
- to encourage the production of local quality foods with a high regional and cultural image
- to support professional agricultural education and training
- to maintain the visual amenity of the rural landscape
- to ensure that rural communities remain vibrant and active
- to generate and maintain employment
- to safeguard and stabilise farmers' earnings.

All of these aims are both honourable and desirable in the agricultural sector. The degree of success of these policies is different in each country, but it is still too early to make firm judgements. When ten new – and predominantly agricultural – members join the EU in 2004 or soon after, the development of the rural sector will put much greater strains on the funds and the resolve of the EU.

Part of the EU's rural development programme is to move with the shift away from agriculture within the rural economy rather than just against it. These policies are in the long term more important than the purely agricultural ones, and can be regarded as part of a **sustainable development** plan; they include:

- the development of rural infrastructure to halt the rural exodus
- to encourage diversification, with such things as niche products and micro-enterprises
- to encourage new enterprises to be based on integrated development and a 'bottom-up' approach
- to protect the rural heritage, buildings and landscapes
- to promote green tourism in rural areas
- to protect natural resources and invest in renewable energy
- to develop new cultural, social and economic services
- to develop better socio-economic links between rural and urban areas.

CASE STUDY: RURAL DEVELOPMENT PROGRAMMES IN LOMBARDY, ITALY

Lombardy is a large and diverse region in NW Italy. With a total population of just under 9 million, it is Italy's most populous *regione* and over 2 million of its inhabitants live in and around Milan, its capital city. Only 23% of the region's population is rural, and a mere 5% of the workforce is engaged in agriculture. The region has two main types of farmland, the flat, low-lying and very fertile Pianura Padana or Northern Plain and the mountainous areas of the Alps to the north and a much smaller area within the Apennines to the south.

The Lombardy regional authority classifies farms into three broad categories:

- The 'professional' farms of the plains, which are geared to large-scale production, are highly competitive and have well-organised marketing outlets through the supermarket chains. The problems of these farms are both a tendency towards **monoculture** of such products as milk or cereals and the posing of great pressures upon the natural environment with the intensive use of fertilisers and pesticides.
- The smaller scale farms of the mountains, which tend to produce high-quality products such as wines, cheeses and cured meats for niche markets. Much of this land is marginal in comparison with the agribusiness farmland of the plains, infrastructure is often poor and the levels of skill generally limited. On the other hand, as these farms are located in highly attractive landscapes, they can easily diversify their incomes through tourism.
- The farms on the peripheries of the cities, which are well organised for marketing and have a wide variety of produce, particularly in the horticultural sector. These farms tend to suffer from fragmentation and the incursion of forms of urban land use.

Over €805 million are being invested in the Lombardy rural development programmes from 2000 to 2006; of this €337 million is coming from EU funding. The regional authority of Lombardy has identified three main priority areas in which the investments are being made:

- Making the agriculture more competitive. This involves investing in making production more efficient, producing higher quality foods, the encouragement of more organic farming and greater diversification, the setting up of new training schemes, and the improving of marketing and technical assistance.

- Improvement of the environment. This sets out to help farmers, particularly in the disadvantaged mountain areas, to reduce the use of polluting substances, to maintain the local landscapes, preserve biodiversity, to increase the amount of land under forestry and to use integrated production methods such as crop rotation.
- Integration of the development of rural areas and rural settlements. This involves the investment of funds in the modernisation of villages, at the same time as preserving their historical character, improving water resources by aqueduct extension, creating better power supplies, and making the rural communities more prepared for emergencies and natural disasters such as landslides and flooding.

CASE STUDY: AIDS IN RURAL SUB-SAHARAN AFRICA: A REAL CRISIS IN THE COUNTRYSIDE

Whatever the problems are in rural Europe, conditions are becoming dire in many parts of sub-Saharan Africa. For a continent with frequent disruptions of the rural economies of many of its countries and consequent displacements of rural populations through drought, flooding, civil wars, tribal conflicts and boundary incursions, Africa is probably now facing its worst crisis in modern times. Of the 36.1 million people in the world who are currently living with HIV/AIDS, 95% are from LEDCs and of these the most badly affected are some 20 countries in sub-Saharan Africa. At present there are estimated to be 25 million people with HIV/AIDS in Africa, a very high proportion of these being part of the rural population. AIDS has killed 7 million agricultural workers in Africa in the last 15 years and is likely to kill another 16 million in the next 20 years. Although initially it appeared that central African countries such as Uganda and Tanzania were the most badly affected by the virus, there has been a recent shift southwards to southern Africa. Botswana, Zimbabwe, Swaziland and Namibia all have between 10% and 26% of their adult populations HIV positive. Of these, it is Botswana that currently has the worst problem with 26% of the adult population infected, 35% of all pregnant women are HIV positive and over 25 000 children have been orphaned as a result of parents dying of AIDS-related conditions. Botswana only has a population of 1.5 million, but was one of the most successful economies in southern Africa because of its stable government and sound economy based upon agriculture and tourism. Now it faces an uncertain future; the country's life expectancy is just 47 but it is likely to fall below 38 by the year 2010.

HIV/AIDS is hitting rural communities in many different ways, and is having a major impact upon nutrition, food security, agricultural production and the structure of rural societies. A series of interrelated problems have caused HIV/AIDS to get a hold of the populations of sub-Saharan Africa, but the root causes are related to poverty

- many men of working age are forced to leave the land and find alternative work in cities where they encounter prostitutes who are HIV positive
- young girls are often forced into prostitution as a method of earning quick money
- communities may lose their economic backbone of young people of working age
- families affected are often obliged to sell property including agricultural tools to pay for healthcare and medicines
- children may have to give up their schooling for similar reasons
- agricultural skills may be lost through family deaths and not passed on to the children
- migrants to the cities who acquire HIV/AIDS frequently return to their village and families to be looked after
- time devoted to caring for the sick leads to the sacrificing of time for food preparation and hygiene
- malnutrition makes sufferers of HIV/AIDS more susceptible to infections.

Various UN agencies are active in dealing with these problems and setting up aid projects and educational campaigns. As the UN organisation IFAD (International Fund for Agricultural Development) states:

> 1 Disease is the symptom; poverty is the cause of failure to cope and contain. Its spread in Africa has a great deal to do with the dislocation of normal lives, from the weakening of family stemming from migration to seek income against a background of desperate
> 5 poverty and from the poverty-induced consignment of women to commercial sex …

Questions

1. Explain how and why our perception of rural areas has changed through time and examine why this change takes place.
2. Examine the contention that there is a crisis in the British countryside.
3. Outline the various conflicts of land use and conflicts of interest that can be experienced in rural areas.
4. In what ways can rural areas in decline be given a new lease of life?
5. What are the main pressures upon rural areas in highly urbanised societies?

Summary Diagram

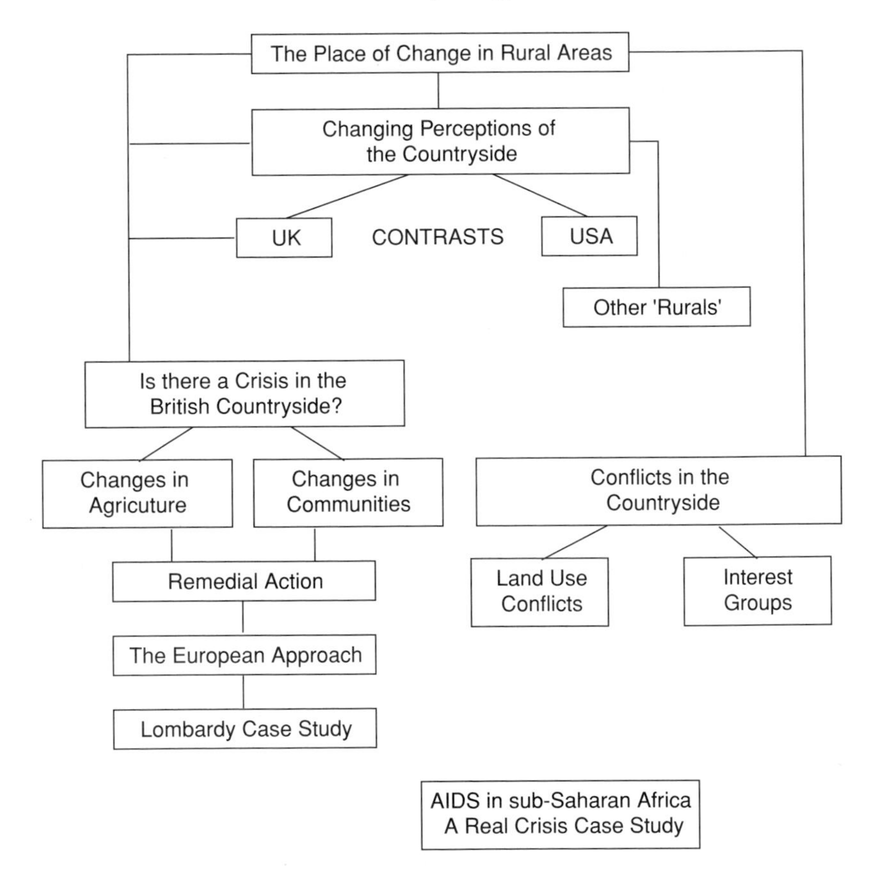

Bibliography

Aston, M. (Ed.), 1988 *Aspects of the Mediaeval Landscape of Somerset.* Taunton: Somerset County Council.

Aston, M., 1985 *Interpreting the Landscape.* London: Routledge.

Bell, D. and Valentine, G., 1995 *Mapping Desire.* London: Routledge.

Bonapace, U. (Ed.), 1977 *Capire L'Italia,* Vol I: *I Paesaggi Umani.* Milano: Touring Club Italiano.

Bollettino della Società Geografica Italiana Ser. XII, Vol. I, No. 4, 1996: Roma.

Bollettino della Società Geografica Italiana Ser. XII, Vol. II, No. 1-2, 1997: Roma.

Cameron, K., 1961 *English Place Names.* London: Batsford.

Carter, H., 1992 *Urban and Rural Settlements.* London: Longman.

Chisholm, M., 1962 *Rural Settlement and Land Use.* London: Hutchinson.

Cloke, P., and Little, J. (Eds), 1997 *Contested Countryside Cultures.* London: Routledge.

Conzen, M. (Ed.), 1990 *The Making of the American Landscape.* London: Harper Collins.

Daniel, P., and Hopkinson, M., 1979 *The Geography of Settlement.* Harlow: Oliver & Boyd.

Denyer, S., 1978 *African Traditional Architecture.* New York: Africana Publishing Company.

Desplanques, H. (Ed.), 1975 *I Paesaggi Rurali Europei.* Perugia: Deputazione di Storia, Patria per L'Umbria.

Everson, J., and Fitzgerald, B., *Settlement Patterns.* London: Longman.

Fraser, D., 1970 *Village Planning in the Primitive World.* London: Studio Vista.

Fabbri, P., 1997 *Natura e Cultura del Paesaggio Agrario.* Milano: CittàStudi Edizioni.

Gilg, A., 1985 *An Introduction to Rural Geography.* London: Edward Arnold.

Guidoni, E., 1975 *Primitive Architecture.* New York: Electra Rizzoli.

Havinden, M., 1981 *The Somerset Landscape.* London: Hodder & Stoughton.

Hill, M., 1971 *The Parliamentary Enclosure of Mendip* (unpublished MSc. thesis).

Hoskins, W.G., 1955 *The Making of the English Landscape.* London: Hodder & Stoughton.

Houston, J., 1963 *A Social Geography of Europe.* London: Unwin.

Ilbery, B. (Ed.), 1998 *The Geography of Rural Change.* London: Longman

Kempe, D., 1988 *Living Underground.* London: The Herbert Press.

Knapp, R., 1992 *Chinese Landscapes: The Village as Place.* Honolulu: University of Hawaii Press.

Khoo, T. S., 1995 *Kemahiran Geografi.* Kuala Lumpur: Pustaka Sistem Pelajaran.

Minca, C. (Ed.), 2001 *Postmodern Geography: Theory and Praxis.* Oxford: Blackwell.

Quercioli, M., 1992 *Le Città Perdute del Lazio.* Rome: Compton Newton Editori.

Parnwell, M., 1993 *Population Movements in the Third World.* London: Routledge.

Pretty, J., 1998 *The Living Land.* London: Earthscan.
Rigg, J. (Ed.), 1996 *Indonesian Heritage,* Vol. 2: *The Human Environment.* Singapore: Editions Didier Millet.
Roberts, B.K., 1987 *The Making of the English Village.* London: Longman.
Roberts, B.K., 1996 *Landscapes of Settlement.* London: Routledge.
Robinson, G., 1990 *Conflict and Change in the Countryside.* London: Belhaven Press.
Schoenauer, N., 1981 *6000 Years of Housing.* New York: WW Norton & Co.
Sestini, A., 1963 *Conosci L'Italia,* Vol. VII: *Il Paesaggio.* Milan: Touring Club Italiano.
Sharp, T., 1946 *The Anatomy of the Village.* Harmondsworth: Penguin.
Turi, E., 1993 *La Civiltà del Villaggio.* Novara: Istituto Geografico de Agostini.
Valentine, G., 2001 *Social Geographies.* Harlow: Prentice Hall.
Walford, N., Everitt, J., and Napton, D. (Eds), 1999 *Reshaping the Countryside.* Wallingford: CABI Publishing.
Yarwood, A., 2002 *Countryside Conflicts.* Sheffield: The Geographical Association.

Websites

www.fao.org (Food and Agricultural Organisation of the UN)
www.ifad.org (International Fund for Agricultural Development (UN))
www.europa.eu.int (EU website)
www.defra.gov.uk
www.countryside.gov.uk (Countryside Agency)
www.environment.gov.uk (Environment Agency)
www.countryside-alliance.org.uk
www.CPRE.org.uk (Council for the Preservation of Rural England, which changed its name and image in June 2003 to the Campaign for Preserving Rural England)
www.soilassociation.org (the major organisation supporting organic farming)
www.inforural.org.uk (information for country dwellers)
www.league.uk.com (the League Against Cruel Sports)
www.english-nature.org.uk
www.nfu.org.uk (The National Farmers' Union)
All UK universities which have Geography Departments also have useful links.

Index

The traditional Dogon village of Sanga on the Bandiagara escarpment in Mali

The hill village of Goriano, located at 720 metres in the Abruzzo region of Italy